Sustainability and Innovation

Sustainability and Innovation

Published Volumes:

Jens Horbach (Ed.)
Indicator Systems for Sustainable Innovation
2005. ISBN 3-7908-1553-5

Bernd Wagner, Stefan Enzler (Eds.)
Material Flow Management
2006. ISBN 3-7908-1591-8

A. Ahrens, A. Braun, A.v. Gleich, K. Heitmann, L. Lißner
Hazardous Chemicals in Products and Processes
2006. ISBN 3-7908-1642-6

Ulrike Grote
Arnab K. Basu
Nancy H. Chau (Editors)

New Frontiers
in Environmental
and Social Labeling

With 18 Figures and 37 Tables

Physica-Verlag

A Springer Company

Prof. Dr. Ulrike Grote
Institute for Environmental Economics and World Trade
Leibniz University of Hannover
μL Koenigsworther Platz 1
30167 Hannover
Germany
grote@iuw.uni-hannover.de

Prof. Dr. Arnab K. Basu
Department of Economics
College of William and Mary
Williamsburg, VA 29187
USA
akbasu@wm.edu

Prof. Dr. Nancy H. Chau
Department of Applied Economics and Management
Cornell University
Ithaca, NY 14853
USA
hyc3@cornell.edu

ISSN 1860-1030
ISBN-10 3-7908-1755-4 Physica-Verlag Heidelberg New York
ISBN-13 978-3-7908-1755-3 Physica-Verlag Heidelberg New York

Physica-Verlag is a part of Springer Science+Business Media

springer.com

© Physica-Verlag Heidelberg 2007

The use of general descriptive names, registered names, trademarks, etc. in this publication does not imply, even in the absence of a specific statement, that such names are exempt from the relevant protective laws and regulations and therefore free for general use.

Production: LE-TEX Jelonek, Schmidt & Vöckler GbR, Leipzig
Cover-design: Erich Kirchner, Heidelberg

SPIN 11811886   88/3100YL – 5 4 3 2 1 0   Printed on acid-free paper

Acknowledgements

Motivated by the need to bring together researchers, practitioners and representatives from international agencies to communicate research findings and to identify areas awaiting future research, an international conference on "The Role of Labeling in the Governance of Global Trade: The Developing Economy Perspective" was held at the Center for Development Research (ZEF) in Bonn, Germany on March 18-19, 2005. This volume contains updated and edited versions of selected papers presented at the conference.

We are grateful to the contributors and the participants at the workshop, namely Tilman Altenburg (German Development Institute, Bonn), Ibrahim Awad (International Labour Organization), Arjun Bedi (Institute of Social Studies (ISS), The Hague), Stefanie Engel (ZEF and ETH Zurich), Eberhard Hauser (Deutsche Gesellschaft für Technische Zusammenarbeit (GTZ), Eschborn), Angelo Lazo (EurepGAP), Vera Scholz (GTZ), Mahesh Sugathan (International Centre for Trade and Sustainable Development, Geneva) and Peter Wobst (Institute for Prospective Technological Studies, European Commission - Joint Research Centre, Seville). Earlier drafts of the papers benefited from reviews by a number of scholars working on related issues: Arjun Bedi (ISS), Indra de Soysa (Norwegian University of Science and Technology, Trondheim), Judith Dean (United States International Trade Commission), Kimberly Elliot (Institute of International Economics, Washington D.C.), Matin Qaim (University of Hohenheim), Andy Thorpe (University of Portsmouth), and Patrick Webb (Tufts University).

Our work in the area of labeling and trade would not have started had it not been for the encouragement and support from Joachim von Braun (ZEF, IFPRI) and subsequently by Klaus Frohberg (ZEF). To them, and a number of others scholars in the field, we owe intellectual debt for comments and discussions on our work over the years: E. Kwan Choi, Kaushik Basu, Ravi Kanbur, M. Ali Khan, Ramon Lopez, Ron Jones and Amy Glass.

Finally, we thank Thi Phuong Hoa Nguyen, Jutta Schmitz and Ellen Pfeiffer for their very able and professional assistance during the workshop. Our gratitude also goes to Benjamin Schraven for his excellent assistance.

Financial support for the conference and towards the publication of this volume was provided by the GTZ and is gratefully acknowledged.

Table of Contents

Outline and Emerging Issues

Ulrike Grote, Arnab Basu and Nancy Chau

The rapid expansion of world trade over the last 50 years has come with a growing recognition that there exist significant cross-border differences in the choice of production techniques. Meanwhile, consumers' concern for conditions of work and product standards also acquires new meaning with the international division of labor made possible through trade. Broadly, these concerns come under two categories: those that relate to environmental performance and food safety, and those that concern labor standards and human rights. In the search for a market-based mechanism that can reconcile these concerns, environmental and social labeling schemes have emerged to serve as valuable sources of information concerning the environmental and social impacts of production processes and methods. Notwithstanding the rapid growth of these labeling initiatives in recent years as a source of consumer information, the supply side impacts of these schemes have also begun to receive attention. In this context, eco-labeling is now widely used to promote environmentally friendly production methods and to ensure food safety standards, while social labels are seen to promote working conditions that are consistent with internationally recognized minimum standards, and that no children have been employed.

Traditional labels emphasized one particular environmental or social aspect of the life cycle of a product, for example, the non-use of a specific input, pesticide or chemical fertilizer. Recent labels tend to follow a more comprehensive and multi-criteria approach. Here, the whole life-cycle of a product including the production and process methods (PPMs) is typically being described. This life-cycle approach takes separately into account production and processing stages, and a variety of types of environmental aspects: resource and energy usage emissions, waste creation or nuisance. In addition to these environmental claims, other process attributes such as animal welfare, biotechnology, packaging, as well as the impact on working conditions and social welfare are increasingly being considered in labeling schemes.

The attractiveness of environmental and social labels stems from their voluntary nature and market-driven approach to achieve environmental and social goals. Through product prices that reflect green and social preferences for consumers, the argument goes that labeling schemes have the potential to realize a shift towards greener and socially conscious production techniques[1]. In contrast to existing trade-related instruments like tariffs, quotas and sanctions, environmental and social labeling takes advantage of green consumerism and the social awareness of consumers and can have the potential to induce voluntary adoption of eco- and socially-friendly production techniques. On the other hand, environmental and social labeling may also give rise to trade repercussions when labeling standards differ across trading partners, and constitute a source of multi-lateral coordination failure. From a policy standpoint, nevertheless, labe-ling has become the preferred instrument for solving high profile trade disputes amongst members of the World Trade Organization (WTO), as evidenced by the tuna-dolphin dispute between Mexico and the United States (US) and the EU-US dispute over the import of hormone-treated beef from the US. This emerging trend makes it even more imperative to take a closer look at the benefits and limitations of product labeling in the governance of global trade.

Starting in the mid 1990s, the initial theoretical work on eco- and social labeling focused primarily on the prospects and problems of eliminating information distortions so that market prices and associated production responses can truly reflect consumer preferences. The potential impact of labeling, therefore, hinges on at least two sets of issues: (i) the size of the price premium that consumers are willing to pay for the attributes adver-tised via labeling, and (ii) consumer and producer receptiveness to the labeling initiatives themselves. In particular, empirical studies carried out in consumer markets of developed countries by Nimon and Beghin (1999), Teisl, Roe and Hicks (2002), and Shams (1995) show that the willingness of consumers in developed countries to pay for labeled products has either been non-existent or too low to support claims that labeling can induce a change in production technology. Subsequent evaluation of the benefits of labeling has raised additional concerns, such as (i) label fatigue, (ii) frau-dulent environmental claims on labels, (iii) labeling-induced unfair compe-tition or green protectionism and (iv) claims of a lack of transparency and rising transaction costs for consumers and producers alike.

A possible reason for this disconnect between earlier theoretical pre-scriptions and empirical observations lies perhaps in modeling labels simply as an instrument that delivers knowledge regarding production methods (and hence eliminates a market distortion on the consumption side), rather than systematically analyzing the *composition* of the label

---

[1] See Basu, Chau and Grote (2003, 2004 and 2006); Basu and Chau (1998; 2001); Mattoo and Singh (1997); Engel (2004); and Bureau, Marette and Schiavina (1998) for details on how voluntary and mandatory labels affect consumer perceptions, production choices and the volume of trade.

itself. Specifically, the issue of optimal label design, in so far as how consumers value the different attributes of a particular label is concerned, remains an open question.

In terms of consumer and producer receptiveness, labeling has provoked an international debate with major policy implications. While there are legitimate reasons for encouraging labeling as a means of improving the environment and protecting human rights, there are also equally important concerns - especially voiced from developing countries - regarding the fairness of these schemes in an international trade context. Indeed, impacts of labeling schemes are complex, depending on the design of and motivations behind such schemes. Especially, private companies or producer associations have recognized labeling to be a useful marketing instrument to improve their image or that of their products, and thus their competitiveness. In some cases, however, labeling might be also abused as a non-tariff trade barrier which aims at protecting the domestic market by making it more difficult for certain products to be imported into a country. In this context, research on the impact of labeling programs on production decisions in the export sector of developing countries, and the credibility of labeling programs in delivering what they promise, have received scant attention in the literature.

This volume showcases research that represents this new frontier of research on the economics of eco- and social labeling. In broad terms, there are two sets of research approaches covered in this volume. The first pertains to consumer and firm level analysis. These studies utilize (i) experimental design and contingent valuation methods to detect the link between label attributes and consumers' willingness, and (ii) household and firm level surveys to gauge the impact of labeling programs on labor supply, production and export decisions in developing countries. The second pertains to the link between labeling and macro-level policy and trade issues. These studies examine the (i) impact of labeling on the volume and terms of international trade between developed and developing countries, (ii) the reverse causal relationship going from openness and other macro-level economic indicators to the incentives to adopt labeling initiatives, and (iii) the policy debate concerning labeling in the international arena.

Debates surrounding the effectiveness of labeling inevitably start with the question as to whether consumers are indeed willing to pay a price premium for 'better' information conveyed through a label. Thus, the first part of the volume begins with two studies of eco-marketing in the US because of the presumption that high-income consumer markets would be a likely location where preferences exhibiting a willingness to pay for information indeed exist. Both these papers focus on two important concerns that can overshadow the potential of labeling programs, namely: (i) whether consumers are indeed willing to pay extra for some of the attributes, and in turn, (ii) whether the labeling of such attributes will make a difference to consumption behavior.

Mario Teisl, Caroline Lundquist Noblet and Jonathan Rubin take a stab at the question of the prevalence of a positive willingness to pay for environmentally friendly passenger vehicles in the state of Maine in the US. The study is based on a random survey of registered vehicle owners in Maine in May 2004. Two experiments were carried out. The first is designed to ascertain the determinants of consumers' assessment of the eco-friendliness of a product. The second experiment is designed to examine consumers' purchase decision, depending on the importance that consumers place on eco-information. Thus, the study takes our under-standing of the intricacies of eco-marketing beyond a simple "yes" or "no" to the issue of willingness to pay, and delve instead into a variety of possible consumer motivations behind the purchase of a product labeled as environmentally friendly. Interestingly, in terms of label design, Teisl, Noblet and Rubin find that more information *need not* be associated with higher eco-rating by consumers. These findings suggest that dual concerns surroundding labeling programs, (i) providing detailed accurate information and (ii) promoting eco-purchases, may not always go hand in hand.

Robert Hicks investigates consumers' preferences for 'Fair Trade' labeled coffee through an experimental survey of consumers in the US. Drawing from the contingent valuation literature to estimate consumers' willingness to pay for public goods, Hicks shows that when the benefits from buying labeled products are public in nature, then information on the existing stock of public goods leads to a higher willingness to pay compared to a label that only conveys product's approved production practices or methods. Using stated preference discrete choice methods, he empirically investigates the impact of information on consumer demand for labeled products and shows that people are willing to pay more as the level of public goods provision increases.

Turning to the opposite end of the supply chain, it has been frequently argued that labeling programs can provide appropriate price incentives for producers who choose to practice environmentally sound production methods or improve labor standards. Meanwhile, labeling may also have negative impacts, particularly for those producers who are faced with binding technological or cost constraints. The latter is particularly relevant for developing country producers, for whom the fruits of globalization may be hindered by green protectionism, and / or labor standard requirements that are inconsistent with current practices.

In this context, Sayan Chakrabarty and Cristina Carambas respectively study at the household level the impact of labeling programs. These are empirical studies based on household surveys, and constitute first-of-its kind to address directly the issue of labeling programs. Chakrabarty presents results of a survey in Nepal conducted to examine the effectiveness of non governmental organizations (NGOs), for example RUGMARK, in reducing the incidence of child labor in the carpet industry. Chakrabarty's data was obtained from interviews with 410 households of Kathmandu Valley in Nepal. Testing and estimating the effectiveness via multiple logistic regression shows that the probability of child labor decreases if the carpet industry has implemented a labeling program, decreases with an increase in adult (household) income, decreases if the head of the household is educated, increases with the age of the head of the household and increases in the presence of more children (aged 5-14) within the household.

Carambas studies the impact of labeling organic rice by drawing on a survey of 123 farm households in Thailand. She uses a cost-benefit analysis to show that although the rice yields of organic farmers are generally lower compared with conventional rice farmers, a positive price premium is nevertheless achieved through labeling so that the net revenues for eco-labeled rice farmers are relatively higher. Her results of the profit distribution analysis reveal that profits for eco-labeled rice both at the farm and export levels are generally higher than profits for conventionally produced, non-labeled rice. In addition, she detected some health benefits for farmers who adopted organic rice production techniques, and showed that Thai farmers are more likely to adopt environmentally friendly production techniques when information about labeling programs is made available to them.

Each of the aforementioned papers takes labeling programs as exogenously given, and looks at consumer and producer responses and consequences on the two sides of the supply chain, respectively. The next paper turns this question on its head, and asks instead: are there systematic reasons behind why countries adopt eco-labeling programs? In addition, do countries that adopt voluntary labeling programs behave as though it pays to do so, and perhaps more importantly, are countries strategically interdependent in their decision to adopt? Arnab Basu, Nancy Chau and Ulrike Grote study these questions based on a data set of national-level eco-labeling programs pertaining to the food industry in 66 developed and developing countries. This study is made up of two parts. It begins with a general equilibrium theoretical analysis of the decision to, or not to adopt. It then turns to an econometric survival analysis of the time to adopt eco-labeling programs. This study reopens the question of a trade and environment linkage, where labeling serves as the signal that links consumer preferences for eco-purchases and producer decisions. Their findings suggest that indeed, export orientation is associated with a higher likelihood to adopt eco-labeling programs. Interestingly, their findings concerning the time pattern of the adoption of an eco-labeling program are consistent with strategic complementarities, and a race to the top.

In a number of high profile trade disputes, labeling programs have been advocated in place of outright trade bans and import restrictions, precisely because it provides the missing informational link between final consumers and producers. These include the dolphin-tuna disputes, the case of the sale of tropical timber, and the presence of aflatoxins in food products. An important question that arises, therefore, is whether labeling programs is a cure-all policy option, particularly in trade disputes concerning hidden product attributes The paper by David Orden and Everett Peterson concerns precisely this question, and singles out in particular an important case in point wherein feasibility of labeling is limited at best. The case in question concerns the longstanding import ban on Mexican avocados by the United States. The rationale for such a ban since its onset in 1914, has been the lack of control on host-specific avocado pests prevalent in Mexico but not in the United States, and the possibility of fruit-fly infestation of destination country orchards subsequent to export. The potential for labeling is limited here since the credibility of such labels, and consumer awareness regarding the potential of pest infestation, may both be in question.

Orden and Peterson examine a systems approach to risk management, employed by the USDA over the course of 1991 - 2005 which led to the sequential opening of the US market to Mexican avocado imports. The study also provides a partial equilibrium model, in which the consumer surplus gains and the producer surplus losses, upon introduction of Mexican avocado imports, are ascertained. Their analysis illustrates the complexities of the issues involved when trade expansion is entangled with technical standards and barriers. It also brings in new dimensions in the labeling debate, such as the role of the domestic industry in the policy decision-making process, and the importance of traceability of a product's country of origin. These are issues that await future research.

The second part of the volume focuses on the policy implications of product labeling and standards on the volume of imports for developed countries and on the export performance of developing countries.

Stéphan Marette offers an overview on the impact of labeling on agricultural trade volume by for different kinds of labels. Given the lack of precise data for evaluating the international trade impact due to labeling, he draws extensively on the given and scattered literature and elaborates on French labels in general, and the wine, cheese and poultry markets in specific. Marette provides details on the market shares and price premia related to these labels and discusses the issue of compliance costs. He concludes that the emergence of new labels and markets may lead to competition shifts that impacts domestic markets and may reshape the nature of competition in world markets. Finally, Marette discusses the role of harmonization, mutual recognition and the concept of equivalence in the context of the existing proliferation of labels, and alludes to the importance of consumer education programs and the need for public regulations aimed at avoiding label proliferation.

Ahmed Ghoneim and Ulrike Grote analyze the impact of labor standards on the export performance by drawing on a survey of 83 firms in the textiles and ready-made garments industry in Egypt. According to their econometric results, several variables related to labor standards show a significant effect on the probability of a firm to export more than 50% of its output and exclusively to the West (namely EU and the US). Second, variables which ensure the enforcement of labor standards have a higher explanatory power for the probability of a firm to perform well in exportting than compliance and awareness variables. Third, firms are likely to self-enforce labor standards based on their expectation to improve their market access and the competitiveness of their export products. Thus, the driving forces leading to the implementation of higher labor standards at the firm level are of an economic nature rather than social. And finally, for those firms with a high volume of exports to Arab countries and for smaller firms (both exporting to the West or Arab countries), the effect of standards might lead to export diversification. Labels indicating that no child labor has been involved in the production process were not known to the entrepreneurs in Egypt. In general, the attitude towards labeling is divided, however, with the majority of enterprises applying negative attributes to labeling.

Spencer Henson and Steven Jaffee explore the impact that food safety standards have on the performance of developing countries and explore the responses of developing countries to the enhancement of these increasingly complex food safety standards. Opposed to the often voiced opinion that consider standards as barriers to trade for developing countries, Henson and Jaffee take a different approach by considering standards as catalysts for development in low and middle income countries. Indeed, standards reduce transaction costs, promote consumer confidence in food product safety, and may stimulate capacity building within the public sector. Thus, they may also create a new landscape that, in certain circumstances, can be a basis for the competitive repositioning and enhanced export performance of developing countries. To better understand the strategic options of developing countries to meet these challenges, Henson and Jaffee draw on the concepts of 'exit', 'loyalty' and 'voice' developed by Hirschman. As a result, they point out that the most positive and potentially advantageous strategy combines 'voice', 'proactivity' and 'offensive' orientations. Consequently, the aim of capacity building should be seen as enhancing the scope to implement strategies that are 'offensive', 'proactive' and involve negotiation rather than on conventional problem-solving and coping strategies, often centered on the development of technical infrastructure.

Bettina Rudloff explores the scope and limitations for applying national food safety and labeling regimes in the framework of the WTO. Her analysis draws on the existing WTO database consisting of 373 food-related dispute cases of which 45 refer to the period before 1995 and 328 after 1995. Not only did the total number of cases increase over time, also more and more developing countries have been involved in food disputes as both defending and complaining parties. By analyzing the data, Rudloff found that the scope for implementing stricter national food safety and labeling regimes is very limited since it has to be based on the submission of a risk assessment. National flexibility exists only with respect to the choice of a specific non tariff barrier (NTB) like an import ban, or process controls in case of a dispute. However, she also points at the important fact that for many food safety issues, no standards have been developed so far and that the standard setting process of the Codex Alimentarius Commission is lengthy. In addition, voluntary and private standards as well as labels are gaining increasing relevance and these are not covered by WTO rules.

Although the papers presented in this volume constitute a step towards understanding some of the hitherto unexplored dimensions of eco- and social labeling, there remain a number of open areas of research on the topic. For instance, and to repeat a recurring theme on the advantage of labeling over interventions through eco- and social standards, one open issue is the ability of labels to remove the information distortion on the consumption side through a price premium. However, as Robert Hicks, Mario Teisl, Caroline Lundquist Noblet and Jonathan Rubin have pointed out in this volume, this willingness to pay differs according to how consumers value the stock of a public good, and on how consumers perceive different attributes attached to a label. This raises the issue of 'free-riding' inherent in the provision of labeled public goods. In other words, are there certain lower and upper thresholds of the public stock over which product labeling can legitimately induce a price-premium? If so, how do these thresholds vary across products and across countries for the same product?

Relatedly, the issue of international differences in attitudes towards the valuation of labeled products remains an open question. In particular, studies on the willingness to pay in developing countries are more or less absent in the literature. While it is obvious that many consumers in developing countries cannot afford to pay higher prices for eco-friendly products, there are nevertheless a number of labeled products that are being sold in developing countries. Little is known about the market potential of labeling in these consumer markets, and whether the labeling of products does make a difference in consumption behavior in developing countries.

Second, social or environmental attributes advertised through environmental and social labels are often multi-dimensional. Fair Trade Coffee is an interesting case in point. While some base the labeling of fair trade coffee on forest certification, others certify coffee as bird-friendly, organic, shade-grown or as organic. These different labeling schemes are also expected to have different effects on sustainability with respect to social, environmental and economic aspects. However, the implications of such a fine degree of product differentiation on the size of the market and consumers' willingness to pay for each type, again remains an open question.

Third, labeling criteria are increasingly accompanied by traceability requirements. Traceability introduces a system whereby it is possible to trace and track products across the entire supply chain. While proper labeling of the final product at the end of the food chain is aimed at assuring food safety to consumers through the information conveyed on the label, traceability systems generally go *beyond* this labeling information to include issues of accountability for every stage of the production process. The inclusion of traceability raises the question regarding the distribution of costs to various actors along the production chain, especially in developing countries, as compared to traditional, non-labeled supply chains.

Fourth, many labeling programs have been implemented for a very short period of time and the information on different schemes is fragmented and dispersed. However, as the years of implementation of labeling programs increase and more countries start to take stock of the labeling programs in certain sectors in their countries, better informed research can evolve. Improved data will allow for an examination of relevant questions like: what is the role of policy intervention (e.g. subsidies) to the production of labeled products on the volume and terms-of-trade? How do country-specific governance aspects of labeled products (monitoring intensity, claims of fraudulent labeling) influence consumers' willingness to pay when the country of origin is an additional attribute on product labels? Do regional trade arrangements influence the volume of trade in labeled products?

This field of research is closely linked to the second part of this volume related to policy implications of product labeling on the volume of imports for developed countries and on the export performance of developing countries. In this context, research on the question of whether and to what extent environmental and social labels may be abused as non-tariff barriers to trade is scant in the literature. The perspective of different developing countries is especially relevant in this context as private and voluntary standards, not covered by WTO rules, become increasingly more prevalent.

References

Basu AK, Chau NH (1998) Asymmetric country-of-origin effects on intra-industry trade and the international quality race. Review of Development Economics 2:140-166

Basu AK, Chau NH (2001) Market access rivalry and eco-labeling standards: are eco-labels non-tariff barriers in disguise? Department of Applied Economics and Management Working Paper, Cornell University, WP 2001-15

Basu AK, Chau NH, Grote U (2003) Eco-labeling and stages of development. Review of Development Economics 7:228-247

Basu AK, Chau NH, Grote U (2004) On export rivalry, eco-labeling and the greening of agriculture. Agricultural Economics 31:135-147

Basu AK, Chau NH, Grote U (2006) "Guaranteed Manufactured Without Child Labor": The Economics of Consumer Boycotts, Social Labeling and Trade Sanctions. Review of Development Economics, Forthcoming

Bureau JC, Marette S, Schiavina A (1998) Non-tariff trade barriers and consumers' information: the case of EU-US trade dispute on Beef. European Review of Agricultural Economics 25:437-462

Engel S (2004) Achieving environmental goals in a world of trade and hidden action: the role of trade policies and eco-labeling. Journal of Environmental Economics and Management 48:1122-1145

Mattoo A, Singh HV (1997) Eco-labelling, the environment and international trade. In: Vossenaar R, Zarrilli S, Jha V (eds) Eco-labeling and international trade. Macmillan Press, London.

Nimon W, Beghin J (1999) Are eco-labels valuable? Evidence from the apparel industry. American Journal of Agricultural Economics 81:801-811.

Shams R (1995) Eco-labelling and environmental policy efforts in developing countries. Intereconomics: Review of International Trade and Development 30:143-149

Teisl MF, Roe B, Hicks RL (2002) Can eco-labels tune a market? Evidence from dolphin-safe labeling. Journal of Environmental Economics and Management 43:339-359

The Design of an Eco-Marketing and Labeling Program for Vehicles in Maine[1]

Mario F. Teisl, Caroline Lundquist Noblet and Jonathan Rubin

1 Introduction

The widespread use of eco-labels suggests they are perceived by some as an effective method of altering consumer behavior. Indeed, several stated-preference studies (Anderson 2003; Donovan and Nicholls 2003; Ozanne and Vlosky 2003; O'Brien and Teisl 2004) and a number of market-based studies have documented the potential for eco-labels (Blamey and Bennett 2001; Teisl et al. 2002; Bjørner et al. 2004). Although some industry sectors have adopted eco-labeling to take advantage of specialized product markets and potential product premiums, others have been sceptical about the touted environmental and economic benefits of these approaches.

Given that eco-labeling is not costless[2], certification and labeling programs may not achieve their objectives unless consumers are willing to pay for the underlying improvements in the production practices specified by the program. However, in addition to being willing to pay for eco-labeled products, consumers must also notice, understand and believe the information presented to them by the product manufacturer. Thus, the success of labeling is contingent upon both the characteristics of the consumer and of the labeling program. Here we provide a review of the literature demonstrating some of the individual and label program characteristics that have been hypothesized, or shown, to influence the effectiveness of eco-information. We then present results from a current study testing some of the individual and label program factors as applied to environmentally preferred passenger vehicles.

We focus on the light-duty vehicle market because light-duty vehicles are one of the major sources of carbon dioxide, nitrogen oxide, carbon monoxide, and volatile organic compound emissions in the United States

[1] Funded by the U.S. EPA – Science to Achieve Results (STAR) Program Grant # 83098801.

[2] The costs of labeling are not generally related to the costs of providing the information, per se; it is the costs associated with changes in production practices needed to meet the label standards and the costs needed to directly link production changes to end-products (e.g., chain-of-custody agreements).

(EPA 2004), and because traditional command and control approaches have been difficult to apply.[3] In addition, although there are several studies (e.g., Brownstone et al. 1996a, b; Bunch et al. 1996; Gould and Golob 1998) indicating a demand for 'greener' vehicles, no one has studied whether an eco-information program is effective in altering consumers' attitudes toward, or purchases of, environmentally preferred vehicles.[4] It is, thus, an open question whether informed customer choice in the light-duty market will lead to these outcomes.

2   Literature Review

The purpose here is to contribute to an understanding of how eco-labels and other types of eco-information work. The specific objectives are first, to develop and test a model explaining a person's propensity to buy an environmentally preferred vehicle as a function of their personal characteristics. The second objective is to extend current research efforts looking at the characteristics of the label and how it influences several metrics known to be important to an eco-label's success. In turn, this section reviews the literature related to the specific individual and label factors studied later in the paper.

2.1   Individual Factors Influencing Eco-Buying

Economic theory suggests that demand for a good is a function of a number of factors; one of these being tastes and preferences. Psychologists have developed a more nuanced delineation of what constitutes tastes and preferences; some of these include a person's level of environmental concern, their perceptions of their effectiveness as an eco-consumer, their faith in the eco-behavior of others and their perception that eco-buying entails compromise.

Environmental Concern - The literature suggests a person's general view of the environment will be a significant factor in promoting eco-purchases, but that concerns more specific to the environmental issues related

---

[3]   For example, Congress's recent inability to increase fuel efficiency standards.
[4]   The research presented within this article will focus on the effects that eco-information programs may have on traditional fueled passenger vehicles, and will not address the case of hybrid vehicles. Throughout this article we will refer to 'greener' vehicles or 'environmentally preferred vehicles'. These terms refer to gasoline-powered vehicles classified as low emission by the USEPA.

to the product under consideration will have a greater impact (Grankvist and Biel 2001; Thøgersen 1999). As air pollution is the primary environmental consequence associated with passenger vehicles, one can imagine a high level of concern regarding air pollution may influence a consumer's choice of vehicle. This possibility is strengthened by the work of Henry and Gordon (2003) in studying the affect of a public information campaign on driving behavior. They recognize that an awareness of the link between driving and poor air quality was needed in order to "influence target behaviors", in this case driving.

Perceived Consumer Effectiveness - Thøgersen (1999, 2000a, b) indicates a consumer's attention to eco-labels is influenced by the belief that a consumer, through their purchase choices, is an important part of the solution to environmental problems.[5] Studies also suggest a positive relationship between perceived consumer effectiveness and willingness to purchase environmentally friendly products (Balderjahn 1988; Lee and Holden 1999; Thøgersen 2000a).

Faith in Others - Another component of environmental concern, recently recognized as a separate construct, is faith in others. Bamberg (2003) points to Ajzens's theory of planned behavior where normative expectations of others may be a factor in an individual's behavior. Gould and Golob (1998) indicate the behaviors of others influenced the participants in their study; drivers often felt no personal responsibility for vehicle air pollution because they noted worse offenders (i.e., observing free-ridership leads to a decreased faith in others and to a decrease in own socially beneficial behavior). Stern (2000) suggests that information, such as provided on an eco-label, may activate consumer's environmental norms by highlighting the benefits to self and others.

Perceived Compromise - While the above-mentioned constructs positively influence one's environmental behavior, there are also barriers to environmentally friendly consumption. One such barrier is when individuals hold beliefs that purchasing environmentally preferred goods entails some increased inconvenience, cost or risk, or entails accepting a decrease in product quality (Stern 1999). Thøgersen (2000b) notes that consumers purchase goods for the perceived utility they will obtain and are unlikely to substitute a good they perceive as providing lower utility merely because it is eco-labeled. Additionally, consumers may see buying an eco-labeled item as a risky behavior if they are unfamiliar with the product or the eco-labeling program (Thøgersen 2000a). As vehicles are a relatively large capital expense, the risk associated with an incorrect decision is clearly

---

[5] This construct is also frequently referred to as 'Ascription of Responsibility to Self' (Stern 2000).

high. Thøgersen (2000a) indicates that eco-labeled products become more difficult to sell when the perceived compromise gets larger. In addition, previous studies also indicate that if other characteristics of a good mono-polize a consumer's attention, the role of environmental concern in the decision will be lessened (Thøgersen 1999). One can imagine that *perceived* inferiority may monopolize a consumer's attention and thus decrease the likelihood of buying green.

2.2   Information Program Factors Influencing Eco-Buying

This sub-section focuses on several program attributes that appear to be important in affecting the impact of information policies: the degree to which all firms are required to provide product information (compul-soriness), the degree of information detail presented to consumers (expli-citness), the degree to which information is required to appear in a format that is uniform across products (standardization) and the organization that is seen as providing the information (source).

Compulsoriness - At the extremes, labeling restrictions are either man-datory or voluntary; most eco-labeling programs fall into this latter cate-gory. Voluntary policies often yield an information environment in which consumers lack data concerning key product attributes. As a result, atten-tion has been devoted to the process by which consumers infer a value for missing information or the process by which missing information affects choice (see Lee and Olshavsky 1997 for a recent review of this literature). This research suggests that consumers look at equivalent attributes from other brands (Jacard and Wood 1988; Ross and Creyer 1992), or other attributes of the same product (Johnson and Levin 1985; Ford and Smith 1987). Others suggest that consumers may not infer missing values at all, but merely pay less attention to a product with missing information (Simmons and Lynch 1991). Teisl (2003) finds that a move from voluntary to compulsory labeling does not significantly alter choice behavior as respondents are able to correctly infer the lack of environmental informa-tion on a product signals the product performed relatively poorly on this characteristic.

A related issue is that the availability of labels in the marketplace seems to play a key role in consumers' use of labels (Thøgersen 2000a). As la-beled products become more common they are more likely to be noticed, appear credible, be useful in making cross-product comparisons and may influence some consumers' perceived consumer effectiveness (Thøgersen 2000a). By definition, a compulsory labeling program increases the availa-bility of eco-labeled products.

Explicitness - Here we define two types of labels differentiated by the level of information detail. Eco-seals, such as seals of approval issued by certification programs, communicate little detail concerning attribute values. Only consumers who are intimately familiar with the certification agency and its standards understand the full meaning of the symbol. At the other extreme are disclosure labels that provide the most detailed information including product attributes, and the disclosure typically involves continuous or categorical information about each element.

Consumer scientists have long understood that more information is not always better because of the possibility of information overload (Scammon 1977) and of distraction from more authoritative information sources (Roe et al. 1999). However, increasing the amount of information on an eco-label can significantly increase the credibility of the label (Teisl 2003; Teisl and Roe 2005) and respondents' ability to correctly identify an environmentally friendly product (Teisl and Roe 2005; Teisl, 2003). One measure of the effectiveness of an information disclosure policy is if consumers can adequately rank competing products by key attributes, as such rankings can be an important input into the consumer choice process (Lee and Geistfeld 1998).

Bei and Widdows (1999) explore how disclosure of simple (summary ratings) versus complex (attribute-level ratings) information differentially affects consumers with different levels of experience and involvement in the product decision-making. They find that both simple and more detailed information improved respondent efficiency, but respondents with previous knowledge of the product category benefit more from the more complex information. However, adding summary eco-ratings can actually backfire, leading to *decreases* in the perceived credibility of the label (Teisl 2003). It seems summary ratings can increase the respondents' level of scepticism about the overall information; this type of response has also been observed in other contexts (Levy et al. 1996; Teisl et al. 1999).

Standardization - At one extreme, a labeling policy can require a specific format, where the firm has no discretion over the presentation. Alternatively, the content of the information may be regulated but the firm has discretion over how the information is presented. Studies suggest that standardized displays provide the largest benefit to consumers (Schkade and Kleinmuntz 1994) because they increase the number of products or attributes considered during choice, allowing for more accurate choice decisions (Coupey 1994). However, standardization can also mask differences. For example, Teisl and Roe (2005) found that when respondents view multiple products bearing a standard eco-seal and different prices they assume the eco-characteristics of the products are similar and are not willing to pay a price differential between the two certified products.

However, when respondents view a similar situation with non-standard eco-seals they assume the environmental characteristics of the higher-priced product are better, and at least some of them are willing to pay the higher price.

Source - Thøgersen (2000c) suggests that the success of an eco-labeling program depends on the credibility of the label. The Angus Reid Group (1991) indicates individuals have very different views about the credibility of different sources of environmental information and a number of studies have found that consumers are sceptical of eco-claims on products (see Peattie 1995). Many other studies find that labels provided by independent sources are trusted more than information provided by business/industry (MacKenzie 1991; Enger and Lavik 1995; Schlegelmilch et al. 1996; Ozanne and Vlosky 2003). However, Teisl et al. (2001b) find that most U.S. survey respondents prefer a federal agency to administer and enforce an eco-labeling program. Differences in the perceived credibility of certifying organizations may be due to differences in respondents' familiarity with the organizations (Teisl 2003; Brown et al. 2002; Thøgersen 2000c).

3   Theoretical Model

To provide a modeling framework to measure changes in consumer choice behavior due to changes in eco-labeled product, one first needs to know how perceptions of environmental quality enter an individual's utility function (here defined in terms of a purchase occasion or decision). The utility evaluation can be represented by the indirect utility function[6]

$$V = v \{ E , p, M, I\} \qquad (1)$$

where E denotes a vector of perceived environmentally related assessments for J products (i.e., $E = [E^S_1,...E^S_J]$), p is a corresponding vector of prices and M denotes income. I denotes a vector of individual characteristics (such as environmental perceptions and perceived consumer effectiveness).

The method that extracts and translates environmental information into an assessment of a product's environmental impact can be viewed as a 'household production' process by which an individual combines her prior environmental knowledge (K), cognitive abilities (A), time (T) and the environmental information (S) presented during the evaluation phase of the

---

[6]   This model is similar to those used by Teisl, Bockstael and Levy (2001a) and Teisl, Roe and Hicks (2002)

purchase decision. Thus, we could model the assessment process during the purchase decision as:

$$E_j = f(S_j, K, A, T | \theta) \tag{2}$$

where E_j denotes the (subjectively) assessed environmental impact of purchasing good j given information set S, S_j is the environmental information displayed about product j at the point of purchase (e.g., an eco-label). The objective level of the environmental impact characteristics represented by the information variable S is denoted by θ. For example, if S represents a 'Low Emissions' claim made on a vehicle label, then θ denotes that the driving of that vehicle produces emissions lower than some preset definition. θ is separate from the assessment function because the individual does not observe it at the time of purchase except through the variable S. Although θ may be unobservable to the consumer at the time of the purchase decision, we include it within the discussion to distinguish between the factor that affects consumer decisions, S, and the one that ultimately determines the environmental impact of production, θ.

We can model the individual's utility[7], once a choice is made as:

$$V_1 = v (E_1(S_1, K, A, T), M\text{-}p_1, I) \text{ if } y_1 \text{ is chosen} \tag{3}$$

where E_1 is the assessed environmental impact of product y_1, S_1 represents the environmental information presented on y_1's label and p_1 is the price of y_1. Typically, the researcher cannot observe E_1, or many of its components, directly necessitating use of the reduced form of (3):

$$V_1 = v (S_1, M\text{-}p_1, I) \text{ if } y_1 \text{ is chosen} \tag{4}$$

The reduced form is not unduly limiting given the policy-relevant variable, S_1, is retained.

Under a random-utility framework, there are unobservable components of the utility function; the individual's utility function is treated as random with a given distribution:

$$V_j = v \{S_j, M\text{-}p_j, I\} + \varepsilon_j \tag{5}$$

where ε_i is the unobservable component of the individual's utility function. Therefore, the choice of product y_1 by an individual indicates that the utility associated with y_1 is greater than any of the other alternatives within a choice set. The probability that the individual will choose y_1 is equal to

[7] The utility function is quasi-linear allowing for aggregation across consumers as the marginal utility of money is held constant. It further assumes only one item is purchased during the purchase occasion (a reasonable assumption for vehicle purchases).

the probability that the utility associated with y_1 is greater than the utility of the alternative:

$$Pr\,(y_1) = Pr\,[v_1\,\{S_1, M\text{-}p_1, I\} + \varepsilon_1 > v_j\,\{\,S_j, M\text{-}p_j, I\} + \varepsilon_j\,] \qquad (6)$$

$$\text{for all } j \neq 1$$

The probability of choosing an alternative can then be estimated using one of various dependent variable modeling techniques.

4   Methods

The analysis is based upon a nineteen-page survey used to gather baseline data on the willingness of Maine citizens to purchase environmentally friendly passenger vehicles. This section clarifies the methods employed in collecting the data.

4.1   Sampling and Survey Administration

In May of 2004, we obtained 1,382,735 records from the Maine Bureau of Motor Vehicles; the records represent everyone who registered a vehicle in Maine within the past 12 months. A random sample of 2,000 was generated from the frame with approximately 800 records removed because they were inappropriate or contained incomplete information.[8] The survey was administered as a three-round modified Dillman between June and August. The total number of respondents was 620, with 169 undeliverable, for a response rate of 60 percent. Our respondents are similar to the characteristics of the Maine adult population as measured by the recent U.S census, except in terms of gender. Although our survey respondents are more likely to be male, the proportion of males correctly reflects the underlying percent of males in the vehicle registration data.

---

[8] Records were rejected if the: primary address was outside the state, vehicle was listed as homemade, registration was for a non-passenger vehicle (e.g., utility trailers, snowmobiles, boats) or records did not have a valid vehicle identification number (VIN). Multiple registrations were also removed, as were records of vehicles older than 1985 (these individuals were assumed to be not in the new car market).

4.2 Survey Design

The survey instrument consisted of seven sections with 41 questions. Sections I and II solicited respondents' opinions on air quality in Maine, the relationship between motor vehicles and air pollution and environmental protection in general. Section III asked respondents about their current vehicle, including the type of vehicle and the importance of various attributes considered during the purchase decision; in Section IV respondents were asked about their search and use of environmental information in the vehicle purchase decision. Sections V and VI incorporated an experimental label test (Experiment I) and a vehicle choice experiment (Experiment II), respectively. These two experiments are analyzed in the paper and their design will be discussed separately (below). The final section of the survey, Section VII, collected demographic characteristics.

Experiment I - Respondents were presented with an eco-label with differing formats and information levels (Figure 1). Five different versions of the survey were created and randomly assigned across respondents. This includes a) the base case where only the State of Maine Clean Car label was presented with no additional text or information; b) the State of Maine Clean Car label with a sliding scale comparing the vehicle to the average of all vehicles in the same class of vehicle; c) the State of Maine Clean Car label with a sliding scale comparing the vehicle to the average of all personal vehicles; d) the State of Maine Clean Car label with a sliding scale comparing the vehicle to the average of all vehicles in the same class of vehicle *and* all personal vehicles; and e) the State of Maine Clean Car label with a thermometer scale comparing the vehicle to the average of all personal vehicles. These diverse label systems allow the analysis to look at two factors that affect a label's effectiveness: the amount of information presented and the consistency of presentation.[9]

Respondents were then asked to rate the label on credibility, perceived environmental friendliness of the vehicle, satisfaction with, and importance of, information. All questions concerning the labels used Likert-type ratings scales. For the credibility question the scale runs from 1, which denotes the label was 'not credible', to 5, which denotes the label was 'very credible'. For the environmental ratings question the scale ran from 1, 'not eco-friendly, to 5, 'very eco-friendly'. In the information load equation 1 denotes 'not enough information, 3 denotes 'just enough information' and 5 denotes 'too much information'. In the information importance equation, 1

---

[9]   Note that the information actually provided to respondents was hypothetical; the vehicle ratings in Figure 1 do not necessarily represent an actual vehicle.

denotes 'not at all important', 3 denotes 'somewhat important' and 5 denotes 'very important'.

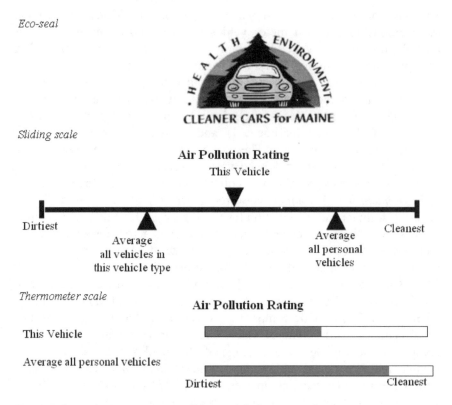

Fig. 1. Information treatments used in eco-label test experiment

Experiment II - Respondents were asked to respond to a two-stage choice scenario; the two stages are designed to reflect the two-stage process of vehicle purchasing (Figure 2) as indicated by focus group participants (Teisl et al. 2004). In the first stage (SI) participants choose a vehicle class (car, van, SUV or truck). After choosing a vehicle class in SI, respondents were then directed to the SII scenario, where they then selected one of three vehicles within their chosen class. Respondents were asked to assume that all vehicles were exactly the same except for the information presented.

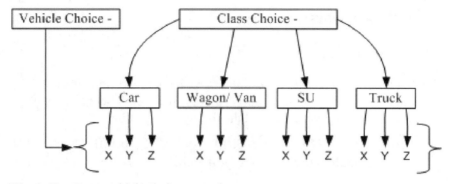

Fig. 2. Two-stage vehicle choice scenario

In SI, respondents are provided with average prices, miles per gallon and scores for criteria pollutants and global warming gases for each of the four classes. The class-level values were generated from two primary sources. Prices for each class were calculated from the National Auto Dealers Association's 'NADA Guides' (NADA.com 2004). The range of class-level fuel efficiency and pollutant scores was calculated based on U.S. EPA's 'Green Vehicle Guide' (EPA 2004). The class-level prices are positively correlated with the criteria pollutant scores (i.e. higher prices are associated with better pollutant scores). Miles per gallon ratings were positively correlated with the global warming scores. The standard deviations of data used to calculate the class averages were used to generate ranges of prices and eco-scores which were randomly assigned to respondents.

In SII, respondents are provided with prices, miles per gallon and scores for criteria pollutants and global warming gases for each of three vehicles. The vehicle-level values were generated from the same sources, and employed the same procedures used to generate the class level values. Respondents were asked to select one of the three vehicles; however respondents were also presented the option of not choosing any of the vehicles presented.[10] If rejection of the choice set was selected, information was then collected on the reason for rejection.

---

[10] Few individuals chose the 'do not choose' option; these observations are not used in the analysis.

5   Data Analysis

5.1   Experiment I

Here we are interested in estimating whether the individuals' eco-assessments of the product differ across the eco-labeling treatments (Equation 2). In turn we estimated the following equation:

$$E_j = \Sigma\alpha_j INT_i + \Sigma\beta_k TREAT_k + \rho_1 CRED + \rho_2 SATIS + \rho_3 IMP + \qquad (7)$$
$$\delta_1 GENDER + \delta_2 AGE + \delta_3 EDUC + e$$

where E_j is the response to the question measuring the individual's assessment of the product's eco-friendliness. INT_j denotes the vector of intercepts ($j = 1 - 3$). $TREAT_k$ denotes the eco-label permutation the respondent viewed (k = A- E). CRED denotes the response to the question measuring the label's perceived credibility. SATIS denotes the response to the question measuring the respondent's level of satisfaction with the information. IMP denotes the response to the question measuring the label's importance. GENDER, AGE and EDUC denote the respondents' gender (1= male; 0 = female), age (in years) and education level (in years). Given the dependent variable is ordered we use ordered-logit techniques. The sign and significance of the β_k provides information on how the reactions of respondents' differed across labels; we test the equivalence of individual pairs of parameters (e.g., $\beta_A = \beta_B$), to determine if the impact of the eco-label are significantly different across the various information treatments.

5.2   Experiment II

In order for eco-labeling initiatives to meet the greatest level of success (i.e. result in the largest number of consumers choosing eco-labeled vehicles), a concrete understanding of the individual characteristics that influence a consumer's reaction to eco-labeling must be established. Here we consider the effect that the personal characteristics of a consumer may have in promoting environmentally responsible purchase behavior in response to eco-labels. Thus, the primary goal of this section is to develop an appropriate empirical model that identifies the variables that influence consumer purchase decisions. The empirical model for any one individual's choice is:

$$\text{Class Choice } [C_j] = \Sigma_j\alpha_j + \Sigma_j\gamma_{1j}USE_1 + \Sigma_j\gamma_{2j}USE_2 + \gamma_3(INC - \qquad (8)$$
$$APP_j - ACD_j) + CRITj*(\gamma_5 + \rho_1 VSUB + \rho_2 FIO + \rho_3 PCE +$$

$$\rho_4 KNOW + \rho_5 CON) + GWG_j *(\gamma_6 + \kappa_1 VSUB + \kappa_2 FIO + \kappa_3 PCE$$
$$+ \kappa_4 KNOW + \kappa_5 CON)$$

$$\text{Vehicle Choice } [C_{k/j}] = \beta_1(INC - APP_k - ACD_k) + CRIT_k *(\beta_2 \qquad (9)$$
$$+ \lambda_1 VSUB + \lambda_2 FIO + \lambda_3 PCE + \lambda_4 KNOW + \lambda_5 CON) + GWG_k$$
$$*(\beta_3 + \tau_1 VSUB + \tau_2 FIO + \tau_3 PCE + \tau_4 KNOW + \tau_5 CON)$$

where Cj and Ck are discrete choice variables indicating an individual's choice of the jth class (either CAR/VAN,[11] SUV or TRUCK) and the kth vehicle (vehicle X, Y or Z), respectively. The class-level intercept terms (α_j) are employed as a means of capturing unobserved class-specific characteristics. USE_1 and USE_2 are constructed to measure the importance that respondents place on specific vehicle-related uses. Specifically, USE_1 measures the average importance (1 = not at all important; 5 = very important) a respondent places on using their vehicle to commute to work and to transport family. USE_2 is a similar measure to quantify the average importance a respondent places on using their vehicle for recreational or work-related hauling. A positive γ_{1CAR} is expected because people who require a vehicle for commuter uses are more likely to choose the CAR/VAN class. We hypothesize respondents who require their vehicle for hauling purposes will most likely choose a TRUCK over a CAR/VAN; this would indicate a negative γ_{2CAR}. We do not hold strong priors on the γ_{jSUV} parameters since SUV's have characteristics that fall in between those of cars and trucks.

Willingness to pay is a function of both price and income. In turn, we create the joint variable $(INC - APP_j - ACD_j)$, where INC denotes the respondents' annual household income.[12] APP denotes the annual cost of purchasing the vehicle and ACD denotes the annual cost of driving. We calculated an annual purchase price for each vehicle provided in the choice scenario (using an interest rate of 6 % and a payment period of five-years). In addition, the annualized vehicle price was adjusted upward by 10 percent to include insurance and tax costs. The annual cost of driving (ACD) variable was created utilizing the formula: ACD = [1/MPG * MILES*

---

[11] Testing indicates the original nesting structure (Figure 1.2) created instability in the parameter estimates and that it was not appropriate to have VAN as a separate nest. Once VAN was combined with the CAR nest, the model became stable.

[12] Since the utility evaluation is at the *individual* level, it is unfortunate we collected *household* income and did not collect household size. As income will be larger, on average, than it should be, the parameter on the joint variable may be underestimated.

CPG* 1.93], where MPG is the miles per gallon stated in the choice scenario for the vehicle, MILES denotes the annual number of miles driven by respondents, CPG is equal to $1.95 - the average cost per gallon of gasoline noted during the time of the survey administration. The last term (1.93) weights the annual gasoline costs to include the annual costs of maintenance.

APP and ACD is intended to capture how ownership and driving costs affect the purchase decision; however, since ACD uses MILES in its construction, ACD could be (at least partially) measuring the individual's need for a vehicle, or need for a larger (more comfortable) vehicle. If true, then one could expect that willingness to pay for a vehicle would increase with increased ACD. We expect this latter effect to be small and anticipate that as a vehicle/class becomes more expensive to own or drive, a respondent will be less likely to choose that vehicle/class. The reasoning behind our assumption is that Maine has a relatively poor public transportation infrastructure and poor weather for much of the year. Thus, we assume most of our respondents need a vehicle due to the lack of substitute means of transportation (i.e., few public transportation alternatives and walking would be uncomfortable for much of the year). Regarding the second possibility for ACD (need for a larger vehicle), we feel that the two USE variables are likely to capture much of this effect.

The parameter estimates on the monetary variables (γ_3 and β_1) should be positive; this would indicate that individuals would be less likely to purchase a vehicle/class with higher relative prices (note: as the annual purchase and driving costs increase, the monetary variable decreases)

CRIT and GWG denote the criteria pollution scores and the global warming scores presented to respondents for each class and vehicle. Both eco-scores were presented on a scale of 1 to 10, where 10 represented the cleanest emission record. It is expected that the coefficients on CRIT and GWG will be positive indicating that higher scores will increase likelihood of purchase.

The interaction variables were created to test whether various personal characteristics influence the importance the respondent places on the eco-information. VSUB, FIO and PCE are variables constructed by using factor analysis on the answers to nine perception questions.[13] The factor analysis indicates that individuals have three underlying factors influencing their responses to these nine questions. Factor one (FIO) reflects a *faith in*

---

[13] Responses to the questions are from a five-point Likert scale where 1 = strongly disagree, 3 = neutral, and 5 = strongly agree. For simplicity we will not fully discuss the factor analysis procedures here - details are available from first author.

others; Factor 2 (PCE) relates to a persons *perceived consumer effectiveness* and Factor 3 (VSUB) measures a person's *perceived compromise* needed when buying a greener vehicles. We hypothesize the parameters on VSUB, FIO and PCE are negative, indeterminate and positive, respectively. If a consumer perceives that an eco-labeled vehicle is not an apt substitute for their normal vehicle, they will be less likely to purchase an eco-labeled vehicle. Consumers with a higher faith in others may be more likely to purchase an eco-labeled vehicle as they feel their pro-environmental choice may be part of a larger effort; however there may also exist an incentive to free-ride and thus the sign on FIO is ambiguous. Consumers with greater perceived consumer effectiveness will be more likely to purchase an eco-labeled vehicle.

KNOW is meant to measure a person's knowledge of vehicles' contribution to air quality; specifically KNOW is a dummy variable where 1 denotes the person feels that all vehicles pollute about the same when driven; 0 otherwise. We hypothesize that the coefficient on the KNOW variable will be negative; individuals who think that all vehicles pollute about the same should place less value on the environmental scores. CON is meant to measure the individuals' general level of concern about the amount of air pollution in Maine (where 1 = not at all concerned and 5 = very concerned). We hypothesize that the coefficient on the CON variable will be positive; individuals who have greater concerns about air quality should place more value on the environmental scores.

Given the two-stage nature of the choice, a nested logit is the most appropriate technique in estimating the results for this data set (Hensher & Greene 2002). Nested-logit models allow for the variances of the random error to be different across groups of alternatives in the utility expressions; this requires scale parameters to be introduced explicitly into the utility expressions (Hensher & Greene 2002). Consistent with the literature, the two scale parameters here are labeled λ (the parameter associated with the class-level utility) and μ (the parameter associated with the vehicle-level utility). To provide consistency with utility maximization, one of the scale parameters must by fixed (typically at 1). Here we estimate the nested-logit model with $\lambda = 1$; this allows the μ's to be free. Give our model contains alternative-specific variables this specification is consistent with utility maximization (Hensher & Greene 2002).

While the existing economic and psychology literatures provide guidance on what explanatory variables should be included in the model, they provide little guidance on whether the variables are important in the class-choice level, at the vehicle-choice level or at both levels in the nesting structure. Given our interest in identifying the form of the model we performed the following analysis on a subset of the data. We first estimated the full model (as presented in equations 8 and 9), then re-estimated the model 1) without any interaction terms; 2) without inter-action terms at the class level only; and 3) without interaction terms at the vehicle level only. Using likelihood ratio tests we can then determine whether inclusion of the additional interaction variables is useful in explaining respondent choices. We also wanted to determine whether the interaction terms were important in explaining differences in individuals' reactions to the criteria pollution scores, the global warming scores or both. Again we used likelihood ratio tests. We find from these analyses that interaction terms are only important at the vehicle level and they are only important in explaining differences in reaction to the criteria pollution scores.[14] The final estimated model is discussed in the results section.

6   Results

6.1   Experiment I

As expected, an increase in the perceived label credibility and in the individuals' satisfaction with the amount of information leads to an increase in the eco-rating (Table 1). Because the regression equation con-tains the information treatment variables, the impact of information quantity on the information credibility and satisfaction ratings is already included. As a result, the label credibility and satisfaction parameters indicate how eco-ratings differ across individuals with different tastes and preferences for, or perceptions of, information, *holding information con-tent constant*. Thus, individuals who are more trusting of, or satisfied with, a given level of information are more likely to view the product as eco-friendly, *ceteris paribus*. Interestingly, individuals with more education provided significantly lower eco-ratings. Gender, age and the stated importance of the information were not significant factors in explaining a respondent's product eco-rating.

In all cases, providing additional quantitative information to the eco-seal leads to *decreases* in the eco-rating of the product; this is consistent with individuals having incorrect priors of a vehicle's cleanliness. One potential measure of the effectiveness of an information policy is if consumers can adequately rank competing products by key attributes when faced with incomplete or imperfect information (see Lee and Olshavsky 1997, for a

---

[14] For brevity we will not fully discuss the analyses here - details available from first author

recent review of this literature). Here, the eco-seal does not provide any explicit environmental score; however, respondents must form some expectation of what the eco-seal means in terms of such a score. The eco-seal by itself apparently led respondents to incorrectly assess the vehicle as being environmentally better than when they were faced with more quantitative information.

Table 1. Regression results for experiment I

Parameter estimates	
Variable name	Coefficient
Intercept	1.825***
Intercept	5.042***
Intercept	6.530***
Treatment A	-4.039***
Treatment B	-4.727***
Treatment C	-5.317***
Treatment D	-4.800***
Treatment E	-5.027***
Perceived credibility	0.403***
Satisfaction with the amount of information	0.483***
Importance of the information	-0.036
Gender	0.093
Age	-0.001
Education	-0.076**

** significant at the five percent level;*** significant at the one percent level
A) only exhibits a State of Maine Clean Car logo with no additional text or information;
B) exhibits a State of Maine Clean Car label with a sliding scale comparing the vehicle to the average of all vehicles in the same class of vehicle;
C) exhibits a State of Maine Clean Car label with a sliding scale comparing the vehicle to the average of all personal vehicles;
D) exhibits a State of Maine Clean Car label with a sliding scale comparing the vehicle to the average of all vehicles in the same class of vehicle and all personal vehicles; and
E) exhibits a State of Maine Clean Car label with a thermometer scale comparing the vehicle to the average of all personal vehicles.

Respondent reactions across label treatments B and C seems to be in the 'correct' direction. That is, respondents gave significantly higher eco-ratings to vehicles environmentally better than a baseline rating (treatment B) compared to those worse than a baseline rating (treatment C). Comparing respondent reactions to treatments C and E indicate the display format of the label (sliding versus thermometer scales) did not impact a respondent's eco-rating of the product.

Comparing treatments B and C with D provides some indication of the importance that respondents place on the different comparative baselines (the same class of vehicle or all vehicles). There is no difference in respondent reactions when they are presented only baseline information about the same vehicle class (treatment B) and when they are presented both the class baseline and the all-vehicle baseline (treatment D). However, there is a significant difference in respondent reactions when they only receive baseline information about all vehicles (treatment C) and when they receive both the class baseline and the all-vehicle baseline (treatment D). This suggests respondents' eco-ratings of vehicle are primarily driven by comparisons between a vehicle and vehicles within the same class. This conforms to previous focus group results (Teisl et al. 2004) where participants indicated that information about the environ-mental friendliness of vehicle should be relative to other vehicles in the same class. Participants reasoned most people shop for a particular class of vehicle because the vehicles within that class better meets their driving needs. They thought it unlikely an eco-label would induce someone to buy a vehicle from differ-rent vehicle class but could induce someone to buy a different vehicle from the same vehicle class.

6.2   Experiment II

The estimated scale parameters (the μ's) lead to Inclusive Values (IV) parameters $(1/\mu)$ that are in the appropriate range $(0 \leq IV \leq 1)$ for a utility maximizing individual (Hunt 2000). Further, the correlation-of-utilities coefficients $(1 - IV^2)$ are relatively close to one (CAR = 0.85; SUV = 0.65 and TRUCK = 0.81) indicating the vehicle alternatives in each class segment are similar to each other (i.e., the nesting structure seems appropriate since the alternatives appear to be reasonable substitutes).

The CAR and SUV-specific variables indicate an individual's use of a vehicle is an important determinant of class choice (Table 2). As commuting becomes more important, respondents are more likely to choose the CAR or SUV class relative to choosing the TRUCK class. Conversely, as hauling becomes more important, respondents are more likely to choose the TRUCK class. The class specific attributes provided in the scenarios had no significant impact on class choice; this may indicate the use characteristics of the class are the primary driver of this choice or that respondents' priors of the different classes are more important than the class-level information we provided them (i.e., the respondent basically ignored the class-level information presented to them).

Table 2. Regression results for experiment II

Variable	Coefficient
Scale parameter (μ)	
CAR	2.610*
SUV	1.613
TRUCK	2.347*
Class choice	
Car-specific variables	
Intercept	-0.315
Importance of commuting (USE1)	0.928***
Importance of hauling (USE2)	-0.988***
SUV-specific variables	
Intercept	-1.513*
Importance of commuting (USE1)	0.692***
Importance of hauling (USE2)	-0.581***
Income – annualized price – annual driving cost (INC-APP-ACD)	-0.096
Criteria pollution score (CRIT)	0.041
Global warming pollution score (GWG)	-0.116
Vehicle Choice	
Income – annualized price – annual driving cost (INC-APP-ACD)	0.165*
Criteria pollution score (CRIT)	-0.006
Global warming pollution score (GWG)	0.098*
Green vehicles are poor substitutes (VSUB * GWG)	-0.005
Faith in others (FIO * GWG)	-0.006
Perceived consumer effectiveness (PCE * GWG)	0.019
All vehicle pollute the same (KNOW * GWG)	-0.014
Concern over air quality (CON * GWG)	0.018

Vehicle choice is positively impacted by the monetary variable; this indicates respondents are less likely to choose a vehicle as the costs of ownership or driving increases. Further, as income increases respondents are less sensitive to the negative price impact. The criteria pollution score is not significant except when its impact is jointly tested with the KNOW variable. The jointly significant negative sign indicates individuals who believe all vehicles pollute about the same when driven are less likely to choose a vehicle having better criteria pollution scores. Although not significant, the signs of the other perception and concern variables are as hypothesized.[15] The positive significant sign on the global warming pollution score indicates individuals are more likely to choose a vehicle displaying a better global warming score. Given all of the GWG interaction terms were deemed unimportant implies that, unlike respondent reactions to the criteria pollutant information, there is no heterogeneity in respondent reactions.

7   Conclusions

In debates surrounding eco-labeling programs, some have argued the lack of consumer response to these products may indicate that consumers do not really care about, or at least are not willing to pay more for, such products. Although this explanation may be valid, it is not necessarily true. One alternative explanation is that consumers do care about and are willing to pay for more environmentally benign products, but the current state of labeling these products is slowing the development of this market. Research in other markets has indicated that well-designed environmental (Bennett and Blamey 2001; Blamey et al. 2001; Teisl et al. 2002; Bjørner et al. 2004) labeling can significantly alter consumer and producer behavior. Experiment II suggests that consumers do value the environmental benefits of more environmentally benign vehicles (at least with respect to global warming gases).[16] Thus, consumer-driven purchases could potentially support an eco-labeled market. A further implication is that consumers who are willing to purchase vehicles with better environmental profiles face a welfare loss (a cost-of-ignorance) when this information is not available (see Teisl et al. 2001a for presentation of this issue).

Experiment I indicates an eco-seal with no other information gave respondents a greener view of the vehicle relative to more quantitative information. This sets up a potential conflict between market dynamics and environmental improvement. A policy of using eco-seals alone would presumeably increase the likelihood of an individual purchasing a labeled vehicle relative to the case of more complete eco-information. This can be seen as follows. Define demand as a function of price (P), income (M) and

---

[15] Given that KNOW is the only interaction term that leads to a significant impact of the CRIT score we used a likelihood ratio test to see whether dropping all of the other interaction terms was indicated. We find that the combination of interaction terms is a significant addition to the model.

[16] Note that the reactions to emissions labeling is directly at odds with current policy reality; in the US most vehicles display criteria emissions labels but no vehicles display global warming gas emissions.

assessed environmental quality (E); where E is a function of the underlying objective level of environmental improvement θ and the label used (S = eco-seal, L = more detailed label). Define θ^1 as a better environmental quality relative to θ^0. Experiment I indicates D (p_1, M, E(S | θ^0)) > D (p_1, M, E(L | θ^0)).

This, in turn, should increase the likelihood of changes in producer behaviors; firms develop new marketing strategies, new eco-products and/or alter the attributes of current products. This would imply the eco-seal alone would lead to a more rapid transition to a more eco-labeled *market*[17] situation (more rapid shifts in demand for, and supply of, eco-friendlier vehicles). However, it is unclear whether the eco-seal alone leads to a more rapid transition to a more eco-friendly *environmental* situation. To see this observe that: D (p_1, M, E(L | θ^1)) > D (p_1, M, E(L | θ^0)).

Hence the relevant comparison is between D (p_1, M, E(S | θ^0)) and D (p_1, M, E(L | θ^1)). Clearly, if D (S | θ^0) ≤ D (L | θ^1) then the more detailed label leads to a more eco-friendly *environmental* situation; however, when D (S | θ^0) > D (L | θ^1) then the result is unclear because it depends upon the differences in vehicle demands and the differences in the θ's. One thing is clear though, consumers who are willing to purchase vehicles with better environmental profiles face a higher welfare loss (a cost-of-ignorance) when this information is provided through the use of eco-seals relative to the label situation (Teisl et al. 2001a).

In reviewing the above conclusions, one should also be mindful of the hypothetical nature of the experiments. First, the market-share dynamics of disclosure policies will be very sensitive to the number of firms in the market and the relative strengths of each firm (see Roe and Sheldon 2002 for an exploration of firm dynamics after the introduction of labeling). Second, using a survey approach may have allowed respondents to evaluate the labels more fully, and with potentially fewer distractions, than they would in an actual purchase setting (see Russell and Clark 1999, for an overview of instances when eco-labels may be less effective in a market setting). Finally, externally validated experiments indicate that when respondents do not face a real budget constraint they are not as sensitive to price differences as they are in real markets.

---

[17] Note we are using a very restrictive definition of market effect. Here we are taking the perspective of someone who defines market success solely in terms of increasing the demand for a labeled product.

References

Anderson R (2003) Do forest certification ecolabels impact consumer behaviour? Presented at CINTRAFOR'S (Centre for International Trade in Forest Products) 20th Annual International Forest Products Market Conference, Seattle WA, 16–17 October

Angus Reid Group (1991) Environment USA

Balderjahn I (1988) Personality variables and environmental attitudes as predicttors of ecologically responsible consumption patterns. Journal of Business Research, 17: 51-56

Bamberg S (2003) How does environmental concern influence specific environmentally related behaviours? A new answer to an old question. Journal of Environmental Psychology 23: 21-32

Bei LT, Widdows R (1999), 'Product knowledge and product involvement as moderators of the effects of information on purchase decisions: a case study using the perfect information frontier approach'. Journal of Consumer Affairs, 33:165–86

Bjørner TB, Hansen L, Russell CS (2004) Environmental labelling and consumers' choice – an empirical analysis of the effect of the Nordic Swan. Journal of Environmental Economics and Management 47:411–34

Blamey R, Bennett J (2001) Yea-saying and validation of a choice model of green product choice. In: Bennett and Blaney (eds) The Choice Modelling Approach to Environmental Evaluation, Cheltenham, UK and Northampton, MA, USA: Edward Elgar., pp 178-201

Brown AS, Brown LA, Zoccoli SL (2002) Repetition-based credibility enhancement of unfamiliar faces, American Journal of Psychology, 115:199–209

Brownstone D, Bunch DS, Golob TF (1996a) A demand forecasting system for clean-fuel vehicles. Organization for Economic Co-operation and Development (OECD) Towards clean transportation: Fuel efficient and clean motor vehicles. Publications Service, OECD. Paris

Brownstone D, Bunch DS, Golob TF, Ren W (1996b) A vehicle transactions choice model for use in forecasting demand for alternative-fuel vehicles. Research in Transportation Economics. Vol. 4:87-129.

Bunch DS, Brownstone D, Golob TF (1996) A dynamic forecasting system for vehicle markets with clean-fuel vehicles. Hensher DA, King J, Oum TH (eds) World Transport Research. Vol.1:189-203.

Coupey E (1994) Restructuring: constructive processing of information displays in consumer choice, Journal of Consumer Research, 21(1): 83–99.

Donovan GH, Nicholls DL (2003) Estimating consumer willingness to pay a price premium for Alaska secondary wood products. US Department of Agriculture, Forest Service, Pacific Northwest Research Station Research Paper PNW-RP-553, October.

Enger A, Lavik R (1995) Eco-labelling in Norway: consumer knowledge and attitudes. In: Stø E (ed), Sustainable Consumption. Report from the International Conference on Sustainable Consumption, Lysaker (Norway): National Institute for Consumer Research (SIFO), pp 479–502.

EPA (Environmental Protection Agency) (2004) Green vehicle guide. http://www.epa.gov/greenvehicle/

EPA (Environmental Protection Agency) (2005) EPAs personal greenhouse gas calculator   http://yosemite.epa.gov/oar/globalwarming.nsf/content/Resource CenterTools GHGCalculator.html - accessed on Monday 12/05/2005.

Ford GT, Smith RA (1987) Inferential beliefs in consumer evaluations: an assessment of alternative processing strategies. Journal of Consumer Research, 14:363–71.

Gould J, Golob TF (1998) Clean air forever? A longitudinal analysis of opinions about air pollution and electric vehicles. Transportation Research-D: Transport and the Environment, 3: 157-169

Grankvist G, Biel A (2001) The importance of beliefs and purchase criteria in the choice of eco-labelled food products. Journal of Environmental Psychology, 21: 405–10.

Henry G, Gordon C (2003) Driving less for better air: impacts of a public information campaign. Journal of Policy Analysis and Management, 22(1): 45-63.

Heschner DA, Greene WH (2002) Specification and estimation of the nested logit model: alternative normalizations. Transportation Research Part B, 36:1-17.

Hunt G (2000) Alternative nested logit model structures and the special case of partial degeneracy Journal of Regional Science 40:89-113.

Jaccard J, Wood G (1988) The effects of incomplete information on the formation of attitudes toward behavioural alternatives. Journal of Personality and Social Psychology, 54(April): 580–91.

Johnson RD, Levin IP (1985) More than meets the eye: the effects of missing information on purchasing evaluation. Journal of Consumer Research, 12(April): 169–77.

Lee DH, Olshavsky RW (1997) Consumers' use of alternative information sources in inference generation: a replication study. Journal of Business Research, 39(3): 257–269.

Lee J, Geistfeld LV (1998) Enhancing consumer choice: are we making appropriate recommendations? Journal of Consumer Affairs, 32(2): 227–51.

Lee and Holden (1999) Understanding the determinants of environmentally conscious behaviour. Psychology and Marketing, 16(5): 373-392.

Levy AS, Derby BM, Roe B (1996) Health claims quantitative study. Briefing presented to the International Food Information Council, US Food and Drug Administration, Washington DC, 8 October.

MacKenzie D (1991) The rise of the green consumer. Consumer Policy Review, 1(2): 68–75.

National Auto Dealers Association (2004) NADA Appraisal Guides, Inc., On-line New Vehicle Price Guide. http://www2.nadaguides.com

O'Brien KA, Teisl MF (2004) Eco-information and its effect on consumer values for environmentally certified forest products. Journal of Forest Economics. 10:75-96.

Ozanne LK, Vlosky RP (2003) Certification from the U.S. Consumer perspective: a comparison of 1995 and 2000. Forest Products Journal, 53(3): 13–21.

Peattie K (1995) Environmental Marketing Management: Meeting the Green Challenge, London: Pitman.

Roe B, Levy AS, Derby BM (1999) The impact of health claims on consumer search and product evaluation outcomes: results from FDA experimental data. Journal of Public Policy and Management, 18(Spring): 89–105.

Roe B, Sheldon IM (2002) The impacts of labelling on trade in goods that may be vertically differentiated according to quality. In: Bohman M, Caswell J, Krissoff B (eds) Global food trade and consumer demand for quality, Boston, MA: Kluwer/Plenum Academic Press.

Ross W, Creyer EH (1992) Making inferences about missing information: the effects of existing information. Journal of Consumer Research, 19(June): 14–25.

Russell CS, Clark CD (1999) The potential effectiveness of the provision of consumer information on product environmental characteristics as a regulatory tool. SØM Meeting in Fredricksdahl, Denmark, 21–23 November.

Scammon DL (1977) Information load and consumers. Journal of Consumer Research, 4(December): 148–55.

Schkade DA, Kleinmuntz DN (1994) Information displays and choice processes: differential effects of organization, form, and sequence. Organizational Behavior and Human Decision Processes, 57(3): 319–37.

Schlegelmilch B, Bohlen GM, Diamantopoulos A (1996) The link between green purchasing decisions and measures of environmental consciousness. European Journal of Marketing, 30(5): 35–55.

Simmons CJ, Lynch Jr. JG (1991) Inference effects without inference making? Effects of missing information on discounting and use of presented information. Journal of Consumer Research, 17(March): 477–91.

Stern PC (1999) Information, incentives and proenvironmental consumer behaviour. Journal of Consumer Policy, 22: 461-478

Stern PC (2000) Toward a coherent theory of environmentally significant behaviour. Journal of Social Issues, 56(3): 407-424

Sutherland RJ, Walsh RG (1985) Effect of distance on the preservation of ware quality. Land Economics, 61(3): 281–91.

Teisl MF (2003) What we may have is a failure to communicate: labelling environmentally certified forest products. Forest Science, 49(5): 13.

Teisl MF, Bockstael NE, Levy AS (2001a) Measuring the welfare effects of nutrition information. American Journal of Agricultural Economics. 83(1): 133-149.

Teisl MF, Peavey S, O'Brien KA (2001b) Environmental certification and labelling of forest products: will it lead to more environmentally benign forestry in Maine? Maine Policy Review, 10(1): 72–8.

Teisl MF, Roe B (2005) Evaluating the factors that impact the effectiveness of eco-labelling programs. In: Krarup S, Russell CS (eds) Environment, information and consumer behaviour, Edward Elgar, Cheltenham, UK, pp 65-90

Teisl MF, Roe B, Levy AS (1999) Eco-certification: why it may not be a field of dreams. American Journal of Agricultural Economics, 81(5): 1066–71.

Teisl MF, Roe B, Hicks RL(2002) Can eco-labels tune a market? Evidence from dolphin-safe labeling. Journal of Environmental Economics and Management, 43:339–59

Teisl MF, Rubin J, Noblet CL, Cayting L, Morrill M, Brown T, Jones S (2004) Designing effective environmental labels for passenger vehicles sales in Maine: results of focus group research. Maine Agricultural and Forest Experiment Station, Miscellaneous Report 434.

Thøgersen J (1999) The ethical consumer. Moral norms and packaging choice. Journal of Consumer Policy, 22:439–60.

Thøgersen J (2000a) Promoting green consumer behaviour with eco-labels. National Academy of Sciences/National Research Council Workshop on Education, Information and Voluntary Measures in Environmental Protection. Washington DC

Thøgersen J. (2000b) Psychological Determinants of Paying Attention to Eco-Labels in Purchase Decisions: Model Development and Multinational Validation. Journal of Consumer Policy, 23: 285-313.

Thøgersen J (2000c) Promoting green consumer behaviour with eco-labels. Workshop on Education, Information and Voluntary Measures in Environmental Protection, National Academy of Science–National Research Council, Washington, DC, 29–30 November.

Performance-based Labeling

Robert L. Hicks

1 Introduction

Many see product labeling as a way for information to impact markets for goods that have a negative social or environmental impact. So long as consumers value not only the good itself but also how the good was produced, it is argued, then a labeling scheme that gives consumers information on the production processes and methods (PPM) will fundamentally alter the market towards greener or socially responsible methods of production. Consumers who value these attributes will be willing to pay more for labeled products. This price premium will provide incentives for producers to choose PPM to mitigate environmental or social problems.[1] Invoking this line of argument, Bass, Markopoulos, and Grah (2000) contend that labeling is "at the heart of many of today's greatest economic, social, environmental, and political challenges, which involve getting the tradeoffs right for sustainable development".

Theoretical studies have attempted to evaluate these claims by investigating the conditions under which eco and social labeling programs can, in fact, "get the tradeoffs right" by allowing consumers to differentiate products according to its associated environmental or social impact. For example, Sedjo and Swaddle (2002) and Basu et al. (2004) investigate the viability of labels and standards in a general equilibrium context for eco-labeled forest products and socio-labels guaranteeing a product was produced without the use of child labor, respectively. In both of these models, equilibrium is based on the price premium an eco or socially conscious consumer is willing to pay to attain a labeled product and the relative costs to the producer of meeting PPM standards.[2] A higher willingness to pay on the part of consumers is seen as a reward by producers in the south and

---

[1] Example of eco-labeling programs includes the dolphin-safe label in the U.S. canned tuna market, the Nordic Swan, and the Blue Angel in Germany. Social labels include the RUGMARK child-labor free rug label began in Germany and in use in the United States, and the FLO and Transfair fair trade label for coffee and other fair trade products.

[2] Basu, Chau, and Grote (2004) model the actions of producers in the north and south, consumers in the north, and the household labor decisions including the

will tend to shift producers toward the eco or socially preferred method of production.[3]

Given the importance of the northern consumer's willingness to pay for the success of labeling programs, empirical studies of the demand for labeled products have shown the existence of premiums for numerous products ranging from canned tuna to organic textiles (Teisl et al. 2002; Nimon and Beghin 1999; Bjorner et al. 2004). However, a closer look at many labeling programs (and producers' opinions about labeling) shows that green PPMs have not been widely adopted and remain a small market segment for most products (Auld et al. 2001; Baharuddin and Simul 1994; Irland and Waffle 2002). This is occurring even while consumers' stated support for eco-labeled products is on the rise. Because of the theoretical importance of the existence of a price premium and the mixed results in the empirical literature concerning the size of the price premium, I investigate consumer preferences for an expanded range of attributes associated with labeled products and show that one explanation for the relatively low willingness to pay for labeled products can be attributed to consumers' lack of information about the performance of labeling programs.

1.1 Performance Labeling

Given the modest price premia found in many studies in the empirical literature, I investigate a labeling approach that goes beyond the traditional labeling paradigm of informing consumers about a good's PPM. Consu-

---

decision to employ child labor. Their model contains numerous testable hypotheses concerning credibility, price premia, relative production costs in the north and south, as well as the role of trade policy for influencing child labor policy. To fully appreciate the impact of performance labeling as presented in this paper, such a general equilibrium approach should be undertaken.

[3] Basu et al. (2004) also discuss label credibility and the monitoring and enforcement of production standards as important determinants of the overall shift in production and associated welfare impacts in the south. A theoretical model by Brown (2001) shows that most of the premia associated with child-free product labels will be captured by the producers in the south and not adult laborers, making households worse off and that labeling credibility will suffer because of false labeling. She concludes "children are found to benefit only if consumers pay an additional amount that can be contributed to a child welfare fund" or bids adult wages in the south to a sufficient level to allow southern households to avoid child labor. Basu (1999) offers a summary of the child labor issue and discusses household production models coupled with a production sector for explaining the child labor decision.

mers may derive value from knowing that the production related to their purchased product met the PPM standard. However, it is also likely that they might be interested in the overall performance of the labeling pro-gram as to how successful it has been in meeting the overall goal set forth by the certifying agency. How the consumer may value the performance of a labeled product is ambiguous. On the one hand, consumers may not be willing to pay for a labeled product if the program is making no appre-ciable difference to the overall problem; while, on the other hand, a well-performing program may be able to capture higher consumer willingness to pay.[4] The performance of a labeling program is collectively defined (e.g. the overall impact of a child-free label) and depends on how the mar-ket for labeled versus non-labeled products work. Compared to a tradi-tional label, in which consumers have no information on a labeling program's overall performance, the performance label offers the consu-mers more information and, perhaps, will increase the price premium asso-ciated with labeled products.

Consider an example from social labeling, namely fair trade labeling of coffee. The goal of the fair trade program is to pay growers an adequate price per pound in order to guarantee the livelihoods of coffee growers in the developing south. However, when purchasing fair trade coffee, the consumer in the north only knows that the product bears a label guaran-teeing a grower a minimum price for their coffee plus a predetermined social price premium.[5] Setting aside the important issue of label credi-bility, the current fair trade label informs the consumer about how her *one-time* purchase of coffee impacted growers. The consumer derives a private benefit from the personal satisfaction of knowing that her purchase en-sured fair wages to farmers. However, the purchase guarantees nothing about meeting the objectives of the labeling program. Important issues like program sustainability, the economic benefits to farmers, and how many farmers participate in this program are not conveyed under traditional labeling programs. To know the full impact of a labeling program, the buyer must also know if the labeling program is meeting the overall goal of the program, a public good determined by collective choice. An indivi-dual's one-time purchase of fair trade coffee provides benefits back to growers, and likely helps in the support of a larger goal related to the labeling program. However, the public benefits - what a purchaser believes

---

[4]   It is also possible that consumers armed with more information on the amount of a public good collectively provided by the labeling program may freeride on the purchases of others.

[5]   The label guarantees the FLO minimum price of $1.21 per pound and pay a social premium of $.05 per pound (Murray et al. 2003 p. 6).

she is contributing to a public good like livelihoods of farmers in the south - are simply unknown to consumers under the current labeling regime making product differentiation across performance attributes impossible.

1.2   Fair Trade Coffee

In this paper, I investigate the impact of including performance attributes on consumer willingness to pay for fair trade coffee.[6,7] I do this because (1) coffee is the fair trade product with the longest history and largest sales volume (James 2000), (2) consumers are used to seeing and evaluating fair trade coffee in the marketplace, and (3) performance metrics are readily identifiable and already measured by fair trade organizations such as Transfair USA.

Following the collapse of the International Coffee Agreement in 1989, real coffee prices fell precipitously to their lowest level in nearly a century while additional countries began producing coffee (e.g. Vietnam). During this time, producers' share of coffee revenues dropped by thirty-five percent as coffee supply increased. In response, the fair trade movement began a labeling campaign aimed at informing consumers that growers received a "fair price" for their product (Transfair USA) and programs were instituted to "facilitate a wider distribution of benefits to small growers" (Taylor 2004). Consumers in the United States and Europe routinely make choices over coffee products that are fair-trade labeled and not. Fair trade coffee in 2003, accounted for only 1% of the world coffee market, yet represented over one-half million growers in the developing

---

[6]   Consider performance labeling in an eco-labeling context. The tuna-dolphin eco-label exhibits significant private and public benefits. A consumer purchasing the eco-labeled product is assured that her purchase of tuna in no way involved the intentional encirclement, capture, or harm to dolphins in the Eastern Tropical Pacific Ocean. That is, consumers benefit from knowing that whatever the status of dolphin stocks in the ocean, their purchasing behavior did not have direct negative impacts on the stocks. It is also possible that over and above these private benefits, consumers may value dolphin stocks directly. That is, their willingness to pay (WTP) for labeled products might vary significantly as a function of dolphin stocks levels. Purchasing the eco-labeled tuna product pro-vides a public good to society (through the preservation of dolphin stocks) even if others in society do not buy dolphin safe tuna. Collectively, consumer's buy-ing the eco-labeled product determine some level of environmental quality that benefits everyone in society.

[7]   For an excellent summary of fair trade coffee and Forest Stewardship Council (FSC) certified timber, see Taylor (2004).

south. In the United States, the fair trade market currently accounts for over 4% of the specialty coffee market and nearly 2% overall (Transfair USA 2005). Finally, fair trade coffee certifying agencies routinely collect performance indicators on the overall achievements of their coffee labeling programs. For example, Transfair reports "Coffee Producer Performance" as the "Additional Farmer Income Generated by Fair Trade in the U.S." and shows that additional revenues have climbed to over twenty five million dollars in 2004. Given farmer participation levels in fair trade programs, rough calculations reveal that farmers can expect to receive no less than almost $70 per year in additional revenues from participating in the program.[8]

Additionally, consumers may also want information about the performance of a labeling program as a further check on label credibility (beyond that of the certifying agency). For example, a recent Wall Street Journal article (Stecklow and White 2004) revealed that only a small portion of the fair trade markup is actually going to coffee growers. Consumers may be quite worried about label veracity - can they believe that the social or environmental claims being made on the label are being delivered? A label that not only informs about the PPM of the product but also relates the performance of the label may in some ways alleviate consumer concerns about whether their price premium is being used to increase producer compensation. For the case of fair trade coffee, a performance metric specifying the increased revenues accruing to program participants would allow consumers to differentiate coffees described in the aforementioned article, where a large portion of the price premium paid by consumers are being captured by the supply chain.

Given that certifying agencies collect and report program performance data (e.g., increases in revenues going to growers, the number of growers enrolled in local fair trade cooperatives), and that such a performance based-labeling initiative could be instituted, several empirical questions need to be addressed to assess the impact of performance-based labeling on the price premium, including (1) does reporting performance as part of the fair trade label always lead to higher consumer willingness to pay for labeled coffee as compared to traditionally labeled coffees, (2) when evaluating a traditional fair trade label, do consumers have an *a priori* belief about program performance, and (3) are performance-based labels always preferred to traditional fair trade labels. Using a stated preference choice experiment, I tackle each of these questions.

---

[8]   It is likely that this figure is a substantial underestimate of revenue increase per farmer, since the number of participating farmers is reported over a five year period, rather than yearly.

In the following section I outline the contingent valuation literature on valuing public goods because similar issues of information provision and the importance of defining the good being valued are central to the contingent valuation methodology. Additionally, I offer a brief introduction to stated preference techniques for measuring consumer preferences for product attributes. Section three outlines the choice experiment including data collection and experimental design. The fourth section details the econometric approach for testing a number of hypotheses concerning fair trade labels, including the importance of certifying agency, and price premia for performance-based fair trade labeling. I conclude with a brief summary of findings and the potential policy implications of performance-based labeling.

2   Literature

A well-known finding in the contingent valuation literature on valuing public goods is the importance of the amount and quality of information for consumer willingness to pay (WTP) for public goods (Mitchell and Carson 1989; Hoehn and Randall 2002). The consumer wants information about the provision rule and the level of public good being purchased in the political market.[9] In the market for labeled goods, information matters in many of the same ways. The consumer buying the product wants to know if it meets the PPM standard, and how the overall market level is impacting the public good. A label that only informs as to the PPM of the product will likely be perceived by consumers to be a very different product to one that meets both the PPM requirement *and* informs the consumer as to the performance of public goods provision.

While more information does inform consumer choice, in a real market place, consumers do not have significant time to devote to studying product labels. In the contingent valuation context, where a large public project is often described, it may be reasonable to assume that voters in a political market would be willing to spend significant amounts of time studying pro-ject information. In a market setting, it is not likely that the average consumer will devote the same amount of time for studying label content.

---

[9] The payment vehicle is also important for contingent valuation experiments. In the eco-labeling setting, the payment vehicle is less important since the consumer buys the green attributes of products through market transactions.

In this study, I employ stated preference techniques to assess how consumers value the public and private components associated with social labeling programs (the technique is termed Stated Preference Discrete Choice (SPDC)). The technique is summarized in Louviere et al. (2000), and has been applied in numerous studies of recreational demand (Hicks 2002); Deshazo and Fermo 2002; Adamowicz et al. 1994) and eco-labeled products (Gudmundssen and Wessels 2000; O'Brien and Teisl 2004).[10] Like contingent valuation, SPDC techniques applied to eco-labeling yield information about preferences by analyzing choices over hypothetical labeled products. Further, SPDC considers a product as a bundle of attributes. Using experimental design techniques, respondents are given product comparisons that are optimal in the sense that they require the respondent to make tradeoffs across the different product characteristics attributes simultaneously.

Additionally, new policy-relevant attributes can be examined; for example, respondents are asked to consider a product under the existing labeing program and one with performance based labeling. Like contingent valuation, SPDC is based upon hypothetical rather than real behavior. There is a growing body of literature comparing revealed and stated preference methods showing that for many cases, parameter esti-ates across revealed and stated preference data are statistically equivalent (Swait et al. 1994, Adamowicz et al. 1994). These tests are seen as validity checks for the SPDC method so that policy guidance resulting from the SPDC model will be relevant for real-world application.

3   Data and Experimental Design

To investigate how information impacts WTP for eco-labeled products, I conduct a split sample experiment. The first sample of respondents are asked to evaluate products based upon the traditional labeling programs informing of production practices and methods only, while the second sample evaluate labeled products with additional information on label performance. In both samples, the PPM of the products and all other aspects of the survey design are identical. Consequently, I am able to isolate the impact of performance-based labeling on the WTP for labeled products.

---

[10] The seafood labeling study of Gudmundssen and Wessels (2000) examines a specific form of performance-based labeling. In their product choice experiment, consumers evaluate products that are either sustainable or not.

I investigate WTP for fair trade labeled coffee because it exhibits significant public goods properties. Buyers of fair trade products are purchasing something like a welfare assistance program in a foreign country for a select group of program participants. The performance of the public good provision due to the label is something that can be enjoyed by everyone in society whether the individual purchases the labeled good or not. Early on in the survey, respondents read the following statement on fair trade products[11]:

Advocates argue that Fair Trade certified products ensure that farmers, workers, and artisans are paid a fair price for their products or labor, don't use child labor or forced labor, have healthy and safe working conditions, use sustainable and environmentally friendly production methods, and have long-term and direct relationships with buyers. Others feel that fair trade is discriminatory against growers and countries that don't have the resources to institute a Fair Trade program.

This statement was purposely worded to convey to respondents, that there are potential up and downsides related to fair trade programs. Participating growers potentially benefit from participation. However, respondents were also informed that for non-participants there might be potential downsides from a fair trade program. I include both perspectives on fair trade because of the need for a balanced survey instrument that give respondents a concise description of the many facets of fair trade, and to lay the groundwork for the performance metrics introduced later in the survey. These metrics are intentionally designed to focus the respondents on the performance of the labeling program for program participants only. Further, respondents are asked several questions about their knowledge of and purchasing habits for fair trade products.

Using a blocked experimental design, I construct two SPDC experiments. Table 1 lists the attributes and levels for each experiment. Note that aside from the two performance metrics, the levels and attributes of the two experiments are identical[12]. Before responding to the choice comparisons shown in Figures 1 and 2 (Appendix), respondents read:

In this section we would like for you to imagine that you are in your favorite campus coffee shop and are looking to purchase a cup of coffee. There are three different brands available for you to purchase. We will ask you to repeat the brand choice several times. Please assume that the brand attributes are identical except for price and any information given on the

[11] The survey is available from the author.
[12] Kenya was dropped from the performance labeling experiment because country of origin effects were found to be small in the traditional labeling experiment and dropping one country of origin increased the design efficiency.

labels. For example, please assume that product quality is the same across the three different brands. If the fair trade or organic label is blank, then there is no information regarding whether that product meets standards or not.

Table 1. Experimental designs

Variable	Traditional label	Performance-based label
Price (Labeled)	{$2.25,$2.50,$2.75,$3.25}	{$2.25,$2.50,$2.75,$3.25}
Price (Non-labeled)	{$1.50,$1.75,$2.00}	{$1.50,$1.75,$2.00}
Country of Origin	{Brazil, Costa Rica, Kenya, Colombia}	{Brazil, Costa Rica, Colombia}
Organic (for non-labeled coffee)	{Yes, No}	{Yes, No}
Certifying Agency	{USDA, Consumer's Union, Coffee Grower's Association}	{USDA, Consumer's Union, Coffee Grower's Association}
Increased Revenue	No Information	{10%, 25%, 50%}
Increased Participation	No Information	{3%, 20%, 40%}

Blocked experimental design techniques were used to select the fifteen sets of 5 questions that maximize respondent tradeoffs across coffees. Although the levels and attributes of the two experiments are identical (except for performance information), the actual levels of the attributes chosen by the experimental design algorithm differ by question, block, and experiment. For each of the two experiments, respondents are randomly assigned to one of the fifteen blocks.

Respondents consisted of students taking large introductory classes (in Economics and Environmental Studies classes) at the College of William and Mary during the fall of 2005. For each treatment, respondents were evenly divided across the economics and environmental studies classes. The survey was filled out during class time and was handed out at the beginning of class. The performance-based experimental design includes two additional attributes. Because the focus of the study was to investigate performance attributes I allocated twice as many respondents to the performance-based design.

4   Model

Consider a consumer faced with a choice over several products. Some of the products are labeled and some are not. The choice problem for the consumer is to choose the best product given preferences and available alternatives. First, consider the choice problem presented in Figure 1, where no performance-based information is given. Let the consumer's indirect utility function for option i be written as

$$V(P_i, \mathbf{C}_i, \mathbf{A}_i, O_i, \varepsilon_i) = \alpha P_i + \mathbf{C}_i' \beta + \mathbf{A}_i' \delta + \omega O_i + \varepsilon_i \qquad (1)$$

where

P_i = Price of Coffee i

$\mathbf{C}_j$ = 1 x k vector of dummy variables indicating country of origin

$\mathbf{A}_j$ = 1 x j vector of dummy variables indicating certifying agency if labeled

O_j = dummy variable = 1 if organic but not fair trade labeled

ε_i = unobserved (by the researcher) error term associated with alternative i

Fig. 1. Traditional label experiment

Notice, that since the label informs about the PPM of the product, the consumer derives some benefit from purchasing the green product (so long as any $\delta>0$). The consumer's choice problem is to choose the product i that maximizes their utility over the choice occasion

$$v_i = \max[V(P_s,\mathbf{C}_s,\mathbf{A}_s,O_s,\varepsilon_s)\forall s \in S] \tag{2}$$

Assuming that the error terms are distributed as GEV I, then the probability of observing the choice of product i can be written as

$$\text{Prob}_i\big(P,C,\mathbf{A},O;(\alpha,\beta,\delta,\omega)\big)$$

$$= \frac{e^{\alpha P_i +C_i{}'\beta+\mathbf{A}_i{}'\delta+\omega O_i}}{\sum_s e^{\alpha P_s +C_s{}'\beta+\mathbf{A}_s{}'\delta+\omega O_s}} \tag{3}$$

Now consider a consumer that faces the choice problem of Figure 2. The consumer is informed of the performance of the labeled product beyond the description of the PPM. Using the performance data, the consumer can gauge how the labeled product is impacting some larger public good through the collective actions of participants in the market. Rewrite the consumer's indirect utility function as

$$V(P_i, \mathbf{C}_i, \mathbf{A}_i, L_i, O_i, \mathbf{G}_i, \varepsilon_i) = \alpha P_i + \mathbf{C}_i'\beta + \mathbf{A}_i'\delta + \theta L_i + \mathbf{G}_i'\phi + \omega O_i + \varepsilon_i \qquad (4)$$

where the definitions of equation (1) are still operative and $\mathbf{G}_i$ is a vector of performance metrics associated with the public good provided by the labeling program. Consumers will choose the optimal product as in equation (2) and from the researcher's perspective, the probability of choosing product i can be written as

$$\text{Prob}_i\left(P, C, \mathbf{A}, \mathbf{G}, O; (\alpha, \beta, \delta, \omega)\right) = \frac{e^{\alpha P_i + C_i'\beta + \mathbf{A}_i'\delta + \mathbf{G}_i'\phi + \omega O_i}}{\sum_s e^{\alpha P_s + C_s'\beta + \mathbf{A}_s'\delta + \mathbf{G}_k'\phi + \omega O_s}} \qquad (5)$$

Comparing equations (3) and (5) reveal the similarities of the choice problem faced by individuals. In both cases, they gain some benefit associated with consuming a good that has been produced with a certified PPM. However, as equation (4) makes clear, consumers are also hypothesized to value the performance of the labeling program with the addition of the term $G_i'\phi$.

The vector δ is capturing several effects. First, it is capturing the effect of certifier credibility and label veracity. Products with more well known and trusted certifiers will likely be preferred to those having either no certification or fly-by-night certification, *ceteris paribus*. Additionally, the consumer may attach the private benefits from purchasing a labeled product and knowing that their purchase had positive impacts on the related public good. If δ is indeed capturing only these effects, then the estimate across the two experiments should be roughly equal given a sufficient sample size. However, it may also be the case that consumers attach priors about a labeling program's performance to the certification agency parameter when performance data is absent. If this is indeed happening, then it is likely that the parameter on certifying agency will play a much larger role under the traditional label than for performance labeled products.

Econometrically, these competing hypotheses can be tested by jointly estimating both models and restricting parameters to be equal across common elements of the choice problem- $\{\alpha, \beta, \omega\}$. This approach assumes, for common data elements, that respondents evaluate information (and make economic tradeoffs) in the same way across the two choice problems. Most importantly, when parameters are restricted across models, the performance-based model simplifies the traditional model when the performance of the labeled product is zero. An alternative estimation strategy freely estimates each set of parameters, and implicitly allows respondents to react differently to labels and information when choosing products.

Since the restricted model is nested within the model where both sets of parameters are freely estimated, we can test whether consumers do in fact value certification veracity and private benefits the same (δ vectors are equal across choice experiments), if they have priors over program performance (δ vectors are not equal) or if they base their purchasing decisions solely on the certification and private benefits associated with the label (ϕ vector is not significant).[13,14]

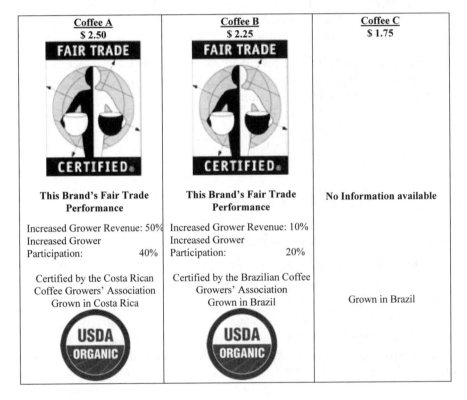

Fig. 2. Performance-based label experiment

---

[13] Previous research has shown that when information is missing, consumers often look for proxies from other attributes of the product or from knowledge about closely related brands (Ross and Creyer 1992; Johnson and Levin 1985; and Ford and Smith 1987).

[14] Unfortunately, my experimental design did not allow me to disentangle the private benefits and certifying agency effects associated with δ. In order to do so, respondents would need to evaluate a subset of labeled products having no certifying agency information.

Using econometric methods proposed by Louviere et al. (2000), it is possible to exploit the differences in equations (3) and (5) to test whether parameter homogeneity holds and therefore to completely isolate how performance impacts valuation of the labeled product. To do this, we estimate two models: a model where common parameters across the two experiments are restricted to be equal, and an unrestricted model where parameters are freely estimated across the two experiments. Define the joint set of parameters from the traditional labeled (denoted by t) and the performance labeled (denoted by p) programs to be estimated as $\psi = \{\alpha^t, \beta^t, \delta^t, \omega^t, \alpha^p, \beta^p, \delta^p, \phi^p, \omega^p, \lambda\}$, where λ is the relative scale parameter that calibrates the restricted parameter estimates to account for error structure differences across the models (see Louviere et al. 2000 for a detailed discussion of the scale parameter).

The likelihood function for the joint model is given by

$$L(\psi) = \sum_{n \in t s \in S^t} \sum y_{sn} \ln\left[\text{Prob}_{ts}\left(P, C, A, O, \left(\lambda\alpha^t, \lambda\beta^t, \lambda\delta^t, \lambda\omega^t\right)\right)\right]$$
$$+ \sum_{n \in ps \in Sp} \sum y_{sn} \ln\left[\text{Prob}_{ts}\left(P, C, A, O, G; \left(\alpha^p, \beta^p, \delta^p, \lambda\omega^p, \phi^p\right)\right)\right] \quad (6)$$

where $y_{sn} = 1$ if respondent n chooses product s. The restricted model can be estimated by setting $\{\alpha^t = \alpha^p, \beta^t = \beta^p, \delta^t = \delta^p, \omega^t = \omega^p\}$. To freely estimate both set of parameters, equation (6) is estimated with only one restriction, $\lambda=1$.[15]

In stated preference studies, where respondents are given all information necessary to make a product choice, Louviere et al. (2000) argue that the error term in the model is capturing the difficulty in assessing and choosing a product. The relative scale parameter (λ) provides a way of measuring the difficulty (commonly referred to as the cognitive burden) of the two experiments (Deshazo and Fermo 2002; Mazzotta and Opaluch 1995; Holmes and Boyle 2005). Given our parameterization of the model, an estimate of λ greater than one reveals that the variance of the error term in the traditional model is smaller than the performance-based model. As the variance of the error term for a given GEV model increases, the unobservable elements of the choice increasingly dominates the discrete choice comparison. Since all information relevant for choice is included in the

---

[15] This is equivalent to separately estimating the traditional and performance-based models.

survey instrument for SPDC experiments, increased dominance of ε indicates increased cognitive burden.

5   Results

Table 2 presents the results from the jointly estimated model and the unrestricted performance-based and traditional labeling model. The results across all three columns in the table reveal striking similarities: in each model a higher price decreases the likelihood of purchasing a given coffee product, and consumers tended to be more willing to purchase USDA certified products. Only in the joint model are country of origin coefficients (relative to Colombia) positive and significant at the five percent level. The organic coefficient on the non-fair trade coffee was not positive or significant (except for the joint model), indicating that consumers choosing the non-fair trade labeled coffee are not more likely to choose a non-fair trade labeled product if it is organic.

The certification agency and private benefit effects of a labeled coffee is always positive and significant for nearly all certifying agencies in both the traditional and performance-based label. The marginal value of a label ensuring fair trade PPM $\left(-\delta/_{\alpha}\right)$ is worth {\$.83,\$1.41,\$.99} for the traditional model (for the Consumer's Union, the USDA, and a foreign country Growers' Association, respectively) and only {\$.17,\$.62,\$.22} for the performance based label. These results suggest that the consumer who is evaluating a traditional labeled product bundles with that label some priors concerning the performance of the labeling program.[16] An analogous explanation is that consumers' who are evaluating performance-based labels are able to evaluate label veracity via the performance data rather than proxying with certifying agency. The performance of the fair trade product was found to significantly increase the likelihood of purchasing a fair trade labeled coffee. Higher performing products are preferred to lower performing products. Both poverty reduction and the level of grower participation had similar effects on the likelihood of choosing the product.

---

[16] An anonymous reviewer conjectured that the respondent may proxy performance with other information on the label such as country of origin when evaluating a traditional label. We tested this conjecture, by estimating a model where country of origin parameters were unrestricted in the choice model. Our results show that country of origin effects were not significantly different across the models.

Table 2. Estimation results

	Parameter	Joint Model	Traditional Model	Performance Model
	Price (α)	-1.0263**	-1.1535**	-1.2413**
		(-7.840)	(-4.608)	(-9.597)
Certifying Agency	Consumer Union (δ)	.3552**	.9596**	.2152
		(2.724)	(2.660)	1.532
	USDA (δ)	.9762**	1.6315**	.7708**
		(8.388)	(5.600)	(5.747)
	Grower's Association (δ)	.5614**	1.1412**	.3373**
		(5.301)	(4.805)	(2.454)
Country of Origin	Brazil (β)	.2582**	.3383	.2624*
		(2.270)	(1.573)	(1.876)
	Kenya (β)	.0713	.1061	N/A
		(.394)	(.498)	
	Costa Rica (β)	.2137**	.1220	.2362*
		(1.979)	(.590)	(1.824)
Label Attributes	Organic (ω)	-.2643**	.0025	-.1981
		(-2.119)	(.9870)	(-1.286)
	Poverty (φ)	1.2127**	N/A	1.9152**
		(4.158)		(6.070)
	Participation (φ)	1.3360**	N/A	1.8763**
		(4.212)		(5.638)
	Relative Scale (λ)	1.0285**	N/A	N/A
		(4.711)		
	Mean Log Likelihood	-1.00737	-1.01240	-0.986745
	N	1270	448	822

Using the joint model I test whether consumers trade-off coffee product attributes in the same way. To estimate the joint model, I restrict the parameters on price, certifying agency, and country of origin across the traditional and performance-based labeling products. I can then investigate consumer priors about the performance of the labeling program. If parameter homogeneity holds (that the restrictions are appropriate) then consumers in both experiments value the certification and private benefit effects in similar ways across the two experiments, and the addition of performance attributes to a traditional label merely increases consumer WTP for labeled products over and above these benefits. Results indicate that parameter homogeneity is rejected using standard log-likelihood ratio tests. This provides evidence that consumers evaluating a traditionally labeled product are willing to pay significantly more for the labeled product than might be expected based on the certification and private benefits effects alone. There is evidence that consumers have priors about the program performance even when no information on performance is provided. Another interesting finding from the joint model is that cognitive burden associated with the performance label seems to be relatively similar to the traditionally labeled product (since $\lambda \approx 1$). Following the interpretation of relative scale parameters, the addition of two additional attributes describing the labeling programs performance makes the choice problem no more difficult than under a traditional label. Contrary to other studies, my results do not show significant increases in cognitive burden when adding label information and provides some evidence that consumers who have no information about performance make guesses as to how effective labeling programs are.[17]

The models can also be used to examine price premiums (or WTP) for eco-labeled products over and above what would have paid for an identical yet not labeled product.[18] Using standard formula for WTP multinomial logit discrete choice models (Hanemann 1999), I show WTP for the traditional label (denoted by the horizontal line) and the performance-based label (assuming a 10% increase in grower participa-tion) for varying levels of increased revenues going to the grower. Notice that in Figure 3 there is a critical value of performance beyond which higher performance increases WTP relative to the traditional label. If the goal of the labeling program is to incentivize grower PPM due to higher consumer WTP, then this finding suggests that a new labeling programs may benefit from starting with a traditional PPM labeling scheme until performance improves beyond a threshold level. As performance rises, the switch to performance-based labeling could begin.[19]

---

[17] Scammon (1997), Roe et al. (1999), and Bei and Widows (1999) explore the issue of quantity of information and its effect on cognitive burden. Of these studies, only Bei and Widows (1999) find that increased information actually improves response efficiency, especially for experienced consumers.

[18] To calculate WTP I compare two coffees: one coffee is not labeled, is priced at $1, and is grown in Colombia, while the other coffee is priced at $1, is labeled, and is certified by the coffee grower's association in the country of origin, also Colombia. To calculate the WTP for performance labeled coffee, I compare the identical coffees except that participation rate increases are 10% and we allow changes in grower revenue to vary for Figure 3. Of course, parameter vectors differ according to the type of label.

[19] An anonymous referee points out that if such a rule were institutionalized, then the rule will likely become internalized in consumer expectations. If this is the case, then a lack of performance data on a label will be a clear signal to consumers that performance criteria are not being met.

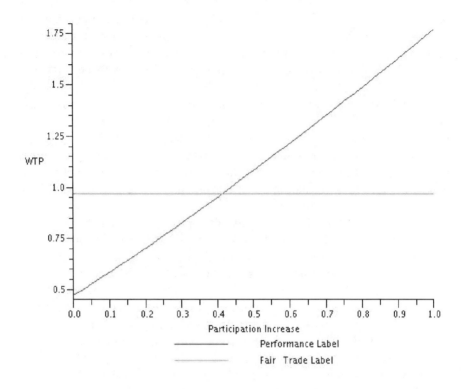

Fig. 3. Price premia for labeled coffee (assumes 10 % increase in grower participation)

6   Conclusion

In this paper, I argue that eco and social labeling schemes as currently implemented leave a lot to the imagination when it comes to consumer preferences for labeled products. Consumers who are interested in more than merely the private benefits associated with purchasing a labeled product are left guessing as to the overall impact of the labeling program on the related public good for the vast majority of labeling programs found around the world today. If consumers' WTP for labeled products is a function of both the overall provision of public goods provided by a labeling program and the private benefits from choosing a labeled product, then the information conveyed by today's labeling schemes may be woefully inadequate from a consumer's standpoint. In this paper, I examine the issue of private and public goods benefits related to a labeling program.

Using two stated preference experiments, I investigate how consumers react to two very closely related purchases of labeled products. In the first experiment, I offer the consumer several coffee products some labeled and some not. The label in the first experiment merely informs the consumer that the production related to the product they purchased did not have negative socio or environmental impacts and is intentionally designed to mimic the majority of labeled products on the market today. In the second experiment, I introduce more information related to the label. In addition to assuring consumers that the product meets socially responsible production standards, the second experiments informs consumers of public goods provided by the labeling program by including on the label, performance metrics of fair trade coffee programs - measured by grower involvement and increased grower revenue.

The results show that consumers are willing to pay more for a higher performing labeled product and provides evidence that consumers' value both public and private benefits from labeled products. Additionally, the econometric specification allows a test of what consumers believe to be the performance of labeling programs when the information is absent from the label. The results show that people probably do have priors over label performance, and, further, labels with more information do not place more cognitive burden on respondents.

Practically speaking, implementation of performance labeling programs does increase the information requirements for certifying agencies, and the results show that poorly performing programs will not receive as high a price premium than a better performing program. Therefore, some care should be taken when starting a new labeling program where poor performance is predictable. The use of performance-based labels does provide further incentives to producers. Since the performance of a labeling program depends on the actions of a number of producers, the label effectively builds a collective reputation of the producer (as indicated by performance). The implications of such a program in a general equilibrium sense is beyond the scope of this paper but is being pursued in other research.[20]

---

[20] It should be noted that the theoretical literature on child-labor free labeling provides interesting hints about the general equilibrium implications of performance-based labeling. Consider the model of Basu et al. (2004), where the developed country production is child-free, and it is competing with labeled and unlabeled products from developing countries. The northern product could be considered a performance-based product having perfect performance and it is competing with traditionally labeled products. We thank an anonymous referee for this insight.

References

Adamowicz WL, Louviere J, Williams M (1994) Combining revealed and stated preference methods for valuing environmental amenities. Journal of Environmental Economics and Management 26:271-292

Auld GB, Cashore B, Newsome D (2001) A look at forest certification through the eyes of the United States wood and paper producers. Paper presented at the Global Initiative and Public Policies: First International Conference on Private Forestry in the 21st Century, Atlanta Georgia, 25-27 March

Baharuddin HJ, Simula M (1994) Certification schemes for all timber and timber products. International Tropical Timber Association, Yokohama

Bass T, Markopoules R, Grah G (2001) Certification's impacts on forests, stakeholders and supply chains: instruments for sustainable private sector forestry services. International Institute for Environment and Development, London

Basu AK, Chau NH, Grote U (2006) Guaranteed manufactured without child labor: the economics of consumer boycotts, social labeling and trade sanctions. Review of Development Economics, forthcoming.

Basu AK (1999) Child labor: cause, consequence, and cure, with remarks on international labor standards. Journal of Economic Literature 37:1083-1119

Bei LT, Widows R (1999) Product knowledge and product involvement as moderators of the effects of information on purchase decisions: a case study using the perfect information frontier approach. Journal of Consumer Affairs 33:165-186

Ben-Akiva M, Morikawa T (1990) Estimation of switching models from revealed preferences and stated intentions. Transportation research 24:485-496

Bjorner TB, Hansen LG, Russell CS (2004) Environmental labeling and consumers' choice - an empirical analysis of the effect of the Nordic swan. Journal of Environmental Economics and Management: forthcoming

Boxall PC, Adamowicz WL, Swait J, Williams M, Louviere J (1996) A comparison of stated preference methods for environmental valuation. Ecological Economics 18:243-253

Brown D (1999) Can consumer product labels deter foreign child labor exploitation? Department of Economics Working Paper 19, Tufts University, Department of Economics, Medford

DeShazo JR, Fermo G (2002) Designing choice sets for stated preference methods: the effects of complexity on choice consistency. Journal of Environmental Economics and Management 44:123-143

Ford GT, Smith RA (1987) Inferential beliefs in consumer valuations: an assessment of alternative processing strategies. Journal of Consumer Research 14:363-371

Gudmundssen E, Wessells CR (2000) Ecolabeling seafood for sustainable production: implications for fisheries management. Marine Resource Economics 15:97-113

Hanemann WM (1999) Welfare analysis with discrete choice models. In: Herriges JA, Kling CL (eds) Valuing recreation and the environment. Edward Elgar Publishers, Northampton

Hicks RL (2002) Stated preference methods for environmental management: recreational summer flounder angling in the northeastern United States. National Marine Fisheries Service, Siver springs, http://rlhick.people.wm.edu/Working_Papers/Stated%20Preference%20Final%20Report_April%2016.pdf

Hoehn, JP, Randall A (2002) The effect of resource quality information on resource injury perception and contingent values. Resource and Energy Economics 24:13-31

Holmes TP, Boyle KJ (2005) Dynamic learning and context-dependence in sequential, attribute-based, stated-preference valuation questions. Land Economics 81:114-126

Irland LC (2002) The elusive green premium. Paper presented at the Sustainable Wood Supply through Market Based Incentives Workshop, Orono, 17 May

James D (2000) Justice and Java: coffee in a fair trade market. NACLA Report to the Americas 34:11-42

Johnson RD, Levin IP (1985) More than meets the eye: the effects of missing information on purchasing evaluation. Journal of Consumer Research 12:169-177

Louviere JJ, Hensher DA, Swait JD (2000) Stated choice methods: analysis and applications with a contribution by Adamowicz W, Cambridge. Cambridge University Press, New York and Melbourne

Mazzotta J, Opaluch JJ (1995) Decision making when choices are complex: a test of Heiner's hypothesis. Land Economics 71:500-515

Mitchell RC, Carson, RT (1989) Using surveys to value public goods: the contingent valuation method resources for the future. RFF Press, Washington DC

Murray DL, Reynolds T, Taylor PL (2003) One cup at a time: poverty alleviation and fair trade coffee in Latin America. The Ford Foundation, New York

Nimon W, Beghin JC (1999) Are eco-labels valuable? Evidence from the apparel industry. American Journal of Agricultural Economics 81:801-811

O'Brien KA, Teisl MF (2004) Eco-information and its effect on consumer values for environmentally certified forest products. Journal of Forest Economics 10:75-96

Roe B, Levey AS, Derby BM (1999) The impact of health claims on consumer search and product evaluation outcomes: results from FDA experimental data. Journal of Public Policy and Management 18:89-105

Ross W, Creyer EH (1992) Making inferences about missing information: the effects of existing information. Journal of Consumer Research 19:14-25

Scammon DL (1977) Information load and consumers. Journal of Consumer Research 4:148-155

Sedjo RA, Swallow SK (2002) Voluntary eco-labeling and the price premium. Land Economics 78:272-284

Stecklow S, White E (2004) At what price virtue? At some retailers, 'fair trade' carries a very high cost stores charge big markups on goods intended to help farmers in poor countries. The Wall Street Journal, June 8:A1

Swait, J, Adamowicz W (1996) The effect of choice environment and task demands on consumer behavior: discriminating between contribution and confusion. Department of Rural Economy, University of Alberta

Swait J, Louviere JJ, Williams M (1994) A sequential approach to exploiting the combined strengths of SP and RP data: application to freight shipper choice. Transportation 21:135-152

Taylor PL (2005) In the market but not of it: fair trade coffee and forest stewardship council certification as market-based social change. World Development 33:129-147

Teisl MF, Roe B, Hicks RL (2002) Can eco-labels tune a market? Evidence from dolphin-safe labeling. Journal of Environmental Economics and Management 43:339-359

TransFair USA (2005) 2005 fair trade facts and figures. TransFair USA, Oakland

USEPA (1993) Status report on the use of environmental labels worldwide. United States Environmental Protection Agency, Washington DC

USEPA (1994) Determinants of effectiveness of environmental certification and labeling programs office of pollution prevention and toxics. United States Environmental Protection Agency, Washington, DC

Waffle R (1997) Forest certification: who profits? Green certification is not needed for sustained forestry. Wood and Wood Products, September:97-101

Do Social Labeling NGOs Have Any Influence on Child Labor?

Sayan Chakrabarty

1   Introduction

In the process of globalization, the labor-intensive industries in South Asian Countries do not only earn a large share of foreign exchange, but also provide a significant share of employment by emphasizing export-led growth. In addition, the growth and expansion of these industries is determined by intra and inter industry competition to gain better comparative advantage across the South Asian Countries. Children are generally fast and quick learners, they do not have any labor union for support, and they are very cheap laborers. Therefore, the opponents of globalization argue that market integration, by increasing labor demand, expands the earnings opportunities of children and thereby inevitably leads to more child labor.

In recent years, the discussion about the impact of globalization on the incidence of child labor has started to evoke a debate in different literatures. Neumayer and de Soysa (2004) argue that countries being more open towards trade and/or having a higher stock of foreign direct investment also have a lower incidence of child labor. They conclude that globalization is associated with less, not more, child labor. Maskus (1997), however, considers globalization as an expanded opportunity to engage in international trade so that a larger export sector will raise the demand for child labor inputs. According to Brown (2002), the rise in the demand for child labor will be accompanied by a rise in the child's wage. This change lowers the return to education and raises the opportunity cost of education, thereby stimulating child labor. On the other hand, Basu and Van (1998) and Basu (2002) argue that any positive income effects that accompany trade openness will help families by meeting or even exceeding the critical adult-wage level at which child labor begins to decline. Contrary to this argument, Edmonds (2002) postulates that increased earning opportunities for parents may change the types of work performed by parents. As a result, children may be forced to take over some of the activities usually performed by adults within their household.

It does not seem to be worth to debate whether changes in local labor markets caused by globalization increases or decreases child labor because no developing country can afford not to participate and/or accept the opportunity of receiving foreign investment by trade creation and trade diversion. However, it might be well argued that the globalization process has been playing a major role in pushing the issue of fair and ethical trade as a priority issue in the international trade debate. That is why the above intellectual debate is very important to address the child labor problem in the international trade literature, especially after the nineties when consumers have learned from the media that a number of the products they purchase could have been produced by child labor.

Therefore, strong concerns throughout the importing countries about the social status of the commodity as well as questions of ethical trade in the globalization process have been raised. India's profits from exporting hand-woven carpets increased from US$ 65 million to US$ 229 million between 1979 and 1983. Due to consumer boycotts that figure dropped to US$ 150 million in 1993, indicating the power consumers have to putting an end to child labor by not buying carpets made by children (Charlé 2001). Activists have been quick in blaming trade liberalization for the negative effects on local labor markets, and have suggested trade sanctions as tools to coerce policy changes aimed at mitigating child labor (Edmonds 2004). Trade intervention has taken the form of either the threat of or the immediate imposition of trade sanctions.

Strong support to the idea of using trade interventions for abolishing child labor arose from the Harkin's Bill, also called the US Child Labor Deterrence Act from 1993. This bill proposed to partially or fully ban the import of goods produced by child laborers. It was based on concerns raised by Senator Harkin about the lack of child protection and the need to ensure mass education (UNICEF 2003). The immediate influence of the bill, which eventually never became law, was dramatic in the case of Bangladesh. Fearing a trade sanction and a loss in market share, almost all child laborers were fired from the garments sector in Bangladesh. An estimated 50,000 children lost their jobs (UNICEF 2003), and nearly 1.5 million families were affected (CUTS 2003). According to UNICEF (2003), 77% of the children, retrenched from the garment industries, were adversely affected in Bangladesh. A majority of the children were pushed into the informal sector, which offers more hazardous and lower paid jobs.

Trade sanctions, thus, have severe limitations. Many doubt the ability of trade sanctions to eliminate child labor (Bhagwati 1995; Maskus 1997). Theoretical models by Maskus (1997) and Melchior (1996) show that trade sanctions or import tariffs against countries where the use of child labor is prevalent do not necessarily reduce the incidence of child labor. On the contrary, the multinational company insisting that its subcontractors fire all child laborers may be doing those children more harm than good (Freeman

1994). After being displaced from the export sector, these children may find themselves worse-off if no viable alternative like education or better working conditions in other sectors exists (Hemmer 1996). In many developing countries, children may also have to work for the economic survival of the family (Grote et al. 1998).

As a result, several measures and initiatives like 'Social Labeling' or 'Codes of Conduct' are directed towards ending the use of child labor. They are increasingly suggested in the context of ethical trade and implemented as an alternative tool to trade sanctions. Social labeling for example acts as a signal in the market informing consumers about the social conditions of production, and assuring them that the item or service they purchase is produced under equitable working conditions (Hilowitz 1997). It is praised as a market-based and voluntary, and therefore more attractive instrument to raise labor standards (Basu et al. 2000).

Many labeling programs have been developed, especially by non governmental organizations (NGOs) like Rugmark, Care & Fair, or STEP. To make sure that these labels remain credible, regular monitoring of the programs is conducted. Generally, if after one or two inspections, children are found working, the licensee is decertified and no longer permitted to use the agency's label. Nevertheless, labeling programs have been criticized on grounds of the credibility of the claims made on their labels. Some organizations believe that credible monitoring is simply an impossible task. For example, the Secretary General of Care & Fair argues that there are "…280,000 looms in India spread over 100,000 square kilometers…" (U.S. Department of Labor 1997, p. 46.). Thus, it is argued that credible monitoring of such a large number of geographically dispersed looms is simply not tenable.

Labeling as a strategy for reducing child labor has received analytical support from Freeman (1994) and Basu et al. (2000) but empirical evidence on this topic is still scarce. Moreover, several recent studies have highlighted the fact that Nepal lacks basic data needed for monitoring employment and labor market conditions.[1] Therefore, this study is an attempt to collect and analyze primary data from Nepali carpet industries. It will focus on the two labeling programs Rugmark and Care & Fair which have been in operation now for 10 years in Nepal. The Rugmark Foundation, established by "Brot fur die Welt", "Misereor", "terre des hommes" and UNICEF in 1995, aims at eliminating the employment of children in the carpet industry by assigning the Rugmark-label to carpets made without child labor. A fund has been set up which is financed by

---

[1] See for instance the report: ILO Nepal Labour Statistics: Review and Recommendations, 1996, Kathmandu.

contributions of the exporting companies. This fund is intended to support the establishment of schools and training institutions in those regions where many children were employed prior to the campaign (Hemmer 1996). Care & Fair is an association established by the German federation of carpet importers. The label does not promise child labor-free products, and monitoring is therefore not needed. It rather supports rehabilitation and education programs for children financed by the imposition of an export charge levied on all carpet imports of member companies to Germany from India, Nepal and Pakistan (Hemmer 1996).

The effectiveness of these labeling programs in eliminating child labor in the Nepali carpet industries will be analyzed in the following. The results of this research will contribute to a better understanding of whether the marketing signals carried by the logos of labeling NGOs are reliable or credible in terms of reducing child labor.

2   Child Labor in Nepal

Nepal is one of the poorest countries in the world with a GNP per capita of US$ 220 and with over half of the population living on less than one dollar a day. The adult illiteracy rate is 60%, and the average household size in Nepal is 5.1 being slightly higher in rural (5.1) than in urban areas (4.8). According to the report on the Nepal Labor Force Survey (NLFS) 1998-99, there are an estimated 3.7 million households in Nepal with a total population of about 19.1 million. The estimated number of Nepalese children under the age of 15 amounts to 7.9 million. Child labor is a widespread problem in Nepal, and can be found with respect to many economic activities. About 500,000 children aged 5 to 9, and 1.5 million children aged 10 to 14 are classified as economically active. This means that their labor force participation rate is 21%, and 61% respectively (NLFS 1998-99).

There are some provisions regarding children in the Nepal Labor Act 2048 (1991). According to the Act, a 'child' is defined as a person who has not attained the age of 14 years (Chapter 1, para. 2). The Act also establishes that "no child shall be engaged in work of any enterprise" (Chapter 2, para. 5). In addition, Nepal ratified the ILO Minimum Age Convention 138[2] in 1997 and the Worst Forms of Child Labor Convention 182 in 2002.

---

[2] ILO Convention concerning minimum age for admission to employment (Convention No. 138), Geneva 1976. See also Ministry of Labor, Main provisions of the constitution of ILO and collection of some of ILO conventions ratified by His Majesty's Government of Nepal, HMG, Nepal, 1997

The Nepali carpet industry is the largest employer and foreign exchange earner in the country. Carpet production in Nepal is concentrated in and around the Kathmandu valley. Nepal's carpet sector experienced its first export boom in 1976. The volume of exports more than doubled within one year increasing from close to 20,000 square meters in 1975 to 47,500 square meters in 1976 (KC 2003). By 1991, this sector contributed to more than 50% of the nation's total exports (Shrestha 1991). The years 1993-94 recorded the highest ever volume of carpet exports, with more than 330,000 square meters amounting to a value of US$ 190 million. By destination of Nepalese carpets, the European market accounts for the biggest share of total export absorption.

After 1994, however, it became internationally well known that the carpet industry intensively employs child laborers - for long hours in any given day. The children work as wool spinners and weavers and some also dye and wash carpets (CUTS 2003). In a study by the Child Workers in the Nepal Concerned Center (CWIN) from the early nineties, 365 carpet factories within the Kathmandu Valley were surveyed, and it was estimated that about 50% of the total 300,000 laborers were children. Of them, almost 8% were below 10 years old, 65% between 11 and 14, and the remaining 27% were between 15 and 16 years (CWIN 1993). A recent study by ILO (2002) estimated that about 7,700 or 12% of the total 64,300 laborers were child laborers in the carpet industries of the Kathmandu valley. According to a survey of 17 carpet factories by the Nepal office of the Asian-American Free Labor Institute (AAFLI), 30% of the workers were found to be less than 14 years of age (CUTS 2003).

However, after hearing about the use of child laborers in the Nepalese carpet sector, consumers in the German market refrained from buying Nepalese carpets (KC 2002). Therefore, from 1995 onwards the carpet sector in Nepal experienced a declining trend in terms of production volume and export earnings. Until the mid nineties, Germany was buying over 80% of Nepali Carpets (Graner 1999) but it then decreased to 64% of the total carpet export from Nepal to Germany (Bajracharya 2004). The decline in the demand for Nepali carpets motivated the government, manufacturers and exporters to participate in the child labor-free labeling schemes. Subsequently, a number of social labeling initiatives such as Rugmark and Care & Fair were introduced in Nepal. The label became a legally binding international trademark in Germany in December 1995, and in 1996 in the US; these are the largest markets for carpet exports from South Asia (CUTS 2003). Currently, almost 70% of the Nepalese carpet industry is licensed by the Rugmark certification system.

3   Removal of Child Laborers and Their Welfare

Only legislation, however sincere it might be in purpose, is unlikely to solve the child labor problem. Since the nature of the problem is rather economic than legal, the labeling NGOs provide schools, health care facilities and hospitals for the displaced child laborers and their family members. If any child is found working by the inspector, then the child and his/her parents or guardians (if available) are interviewed to have complete information about the affordability of the parents to send their child to school. In case of failure or if the children are destitute or orphans, they are taken into the rehabilitation center of Rugmark for long-term reha-bilitation. The rehabilitation center also provides the opportunity for chil-dren to meet with their parents and guardians. These centers have complete hostel facilities, where the children get counseling, medical treatment, recreational activities, etc. Children over 14 years are encouraged to join vocational training programs, which are also financed by labeling NGOs. An emphasis is also put on physical fitness and extra-curricular pursuits such as music and art for the children.

In addition, various supporting programs like school tuition exemption, books, uniforms, and even food are offered by the labeling NGOs to former child laborers and other children of the households of its licensee or subsidiary factory. Thus, they aim at compensating some opportunity cost of child schooling. By the middle of December 2002, the Nepal Rugmark Foundation had removed 478 child laborers from the carpet industries in 40 different districts in Nepal (Rugmark Bulletin 2004).

The NGOs are not only targeting the displaced child from carpet indus-try but also offer welfare programs for other children and adult members of the household who are attached to labeling NGOs. According to the Rugmark Bulletin (2004) "hundreds of children are living in carpet facto-ries and helping their parents or guardians in daily chorus like cooking, taking care of younger babies or just doing nothing. They are already of school going age or older. Some of these children are there without parents but not enrolled in school and some have parents but the parents are econo-mically unable to send their children to school. These children, if not sent to schools, soon join carpet work becoming child laborers". Therefore, to minimize the risk of children to become child laborers in the future, the labeling NGOs have established school, day care cum education centers. Rugmark has supported 11 day care centers covering around 275 children (Rugmark Bulletin 2004).

Labeling NGOs are often giving priority to community-based re-habili-tation. This means that every effort is made to reunite the children with their families, so that they do not become alienated from their communities. Children who return to their families are given for example four levels of support depending upon their need, like support for school fees, books, uniforms and other materials. The foundation has developed and implemented an awareness program for carpet workers in factories under Rugmark license consisting of i) child rights, including education and gender issue, ii) family planning, girl trafficking, HIV AIDS, and iii) health, nutrition, sanitation and working environment. These programs are basically targeted to the carpet workers and their families (Rugmark Bulletin 2004).

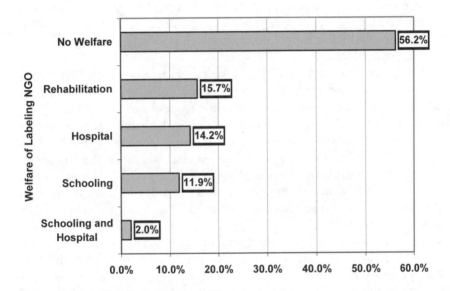

Source: Own Survey (2004).

Fig. 1. Different welfare activities of labeling NGOs

The different welfare activities of labeling NGOs are presented in Figure 1. While 56% of the households are not receiving any welfare activities by labeling NGOs, Figure 1 shows that 16% of the households send their children to rehabilitation centers, 15% are receiving health benefits, 12% are receiving schooling benefits, and 3% are receiving both schooling and health benefits. The rehabilitation center has hostel, food, school, and health benefits for the retrenched children of the carpet industries.

4   The Link Between Social Labeling and Supply of Child Labor

Child labor is demanded by profit-maximizing firms (producing carpet); it is expected that social labeling controls the demand of child labor by registering all the carpet firms within a particular industry and label on the carpet. However, social labeling has also supply side effects on child labor by its rehabilitation and welfare activities. The labeling NGOs like Rugmark and Care & Fair have different programs to educate the members of the household working under the labeling umbrella. The welfare activities like i) sponsorship for education of carpet worker's children, ii) day care centers, iii) awareness programs for carpet workers and their families, and iv) health care facilities might have some direct and/or indirect influence on the child labor supply decision which is mostly taken by the adult members of the family.

By offering families with different welfare programs to send their children to school, one can obviously alter the calculus of their decision making on child labor. The retrenched child has a chance to work in different sectors other than carpet industries or non-labeling carpet industries where labeling NGOs have no control. If social labeling NGOs have no influence on child labor supply reduction, the whole labeling program might be worthless, indicating that social labeling fails to reduce child labor supply rather than shifting child labor from one industry to another; therefore, it has no sustainable/permanent effect on the child labor problem. If the effects of the welfare programs by labeling NGOs are successful the household would not send the child to any work.

The idea of establishing the social labeling NGO is to protect the child labor problem by offering different welfare programs. Therefore, the supply effect of social labeling is very important to address the child labor problem. This study is an attempt to see whether social labeling NGOs have any influence on child labor supply. More precisely this study wants to see whether there is any difference between the decision of labeling and non-labeling households concerning child labor.

5   Data Collection and Methods of Analysis

The main objective of this paper is to identify the effect of social labeling NGOs on the child labor supply decision in Nepal. This study takes into account the determinants of child labor used in various theories mentioned above and considers the influence of social labeling NGOs as a new determinant of the child labor decision by households. In accordance with the ILO convention 138, this study defines children from 5 to 14 years of age who were working in the last two months when this survey was conducted, as 'child laborers', no matter whether the children were working full or part time.

5.1 Survey in Nepal

The data collection in the Kathmandu valley in Nepal was based on primary and some secondary information of households working in the carpet industry. In order to decrease the variances and therefore increase the efficiency of the tests and precision of the estimations, the population was stratified with respect to sources of disturbing heterogeneity. The main suspected sources of heterogeneity were:

1. Administrative differences of regions.
2. Important time points[3].

This study stratifies the population and sample data by equi-proportional sizes with respect to the level of these variables and then draws a simple random sample from each stratum (Levy and Lemeshow 1999). After stratification, the field workers visited carpet industries from the lists of Rugmark and Care & Fair to locate the labeled carpet industries, and visited the non labeled carpet industries from the same area as well.

The major challenge of this study was to locate the stratified households and getting a large enough random sample, so that a reasonable degree of confidence could be reached for statistically significant results. Appendix 1A, 1B, 1D shows sample sizes of different administrative regions at Kathmandu Valley.

There was no base line survey after 1993 that lists the children who lost their job from the carpet industries by the social labeling initiatives but there was a list of the children who were educated by the labeling NGOs schools in different parts of the Kathmandu valley. The other three available lists contain the addresses of the carpet industries provided by CCIA (Central Carpet Industries Association), Rugmark and Care & Fair.

In selecting the sample of carpet industries, the status of its registration by the labeling NGOs was taken into account. So, the sample was stratified by labeling households and non labeling households (see Appendix 1C &

---

[3] The NGOs came into operation in 1995. Therefore, this sampling has to consider whether a present member of a household was a child before 1995 or after 1995. The results shown here consider the second group.

1D). A labeling household is defined as a household with at least one person working in industries registered by labeling NGOs and no member working in any non labeling industry. A non labeling household is a household with at least one person working in the unregistered carpet industry and nobody of the household working in the registered industry.

To compare the situation of labeling and non labeling households, the surveyed households were split into two parts; approximately half of them were selected from labeling and half of them from non labeling households. Appendix 1C shows that the quantitative study covered a total of 1,971 persons in 410 households. 56% of the households were involved with labeling NGOs and 44% were not involved with labeling NGOs.

5.2 Econometric Method

Logistic regression is the most appropriate statistical method to assess the influence of the independent variables on a dichotomous or polytomous dependent variable. A list and description of the dependent and independent variables is to be found in Tables 1 and 2.

We use a binary multiple logistic regression, and define the probability that a child is being employed in the following way:

$$\text{logit}(p) := \ln\left(\frac{p}{1-p}\right) = \alpha + \beta'X \tag{1}$$

where
p   = Probability (Child is employed | X)
α   = Intercept parameter
ß   = Vector of slope parameters
X   = Vector of explanatory variables

Table 1. Variables used for statistical calculation at household level

Variable name (SAS)	Variable Description	Type of the Variable
HH_Id	Household Id	Key
HH_HoH_Age	Age of the Head of Household	Continuous
HH_HoH_Sex	Gender of the Head of the Household	Binary Categorical
HH_HoH_Edu	Education of the Head of the Household	Categorical
HH_Size	Actual total permanent members of the household	Continuous
HH_IncGT14	Last month total income of family members older than 14 (adults)	Continuous
HH_Debts	Actual total outstanding debts incl. interest and costs	Continuous
HH_N_ChildLE14	Total actual number of children (<=14)	Continuous
HH_IsAnybodyInLBLInd	Is anybody of the family working in a labeled industry?	Binary Categorical
HH_IsAbsDolPov	Absolute poverty ($)	Binary Categorical
HH_IsAnyChildLab	Has there any child been working in the household in the last two month full time or part time?[4]	Binary Categorical

---

[4] If the working time per day is eight hours or above, then the child laborer works full time. If the working time per day is at least two and less than eight hours, then the child laborer works part time.

Table 2. Variables used for statistical calculation per child in household

Variable name (SAS)	Variable Description	Type of the Variable
Ind_IsThisChildLab	Has this child (age 5-14) been working in the last two month full time or part time?	Binary Categorical
Ind_NGOAssistChild	Is the child helped by labeling NGO?	Binary Categorical
Ind_Sex	Gender of the child	Binary Categorical
Ind_Mother's_Job	Mother's Job of the child	Categorical

The null hypothesis is $\beta_i = 0$ for all i. We divided the explanatory variables into two sets: variables describing household characteristics and variables describing each individual child of a household. This procedure will lead to two approaches: in the first sub-model (2), we only concentrate on household characteristics as explanatory variables (X_H) (see Table 1) and determine the probability that at least one child in a household is employed (see definition above).

$$\text{logit}(p_H) := \ln\left(\frac{p_H}{1-p_H}\right) = \alpha + \beta'X_H \tag{2}$$

where p_H = Probability (HH_IsAnyChildLab | X_H)

In the second sub-model (3), we are interested in the probability of an individual child to work. In this case, household and individual characteristics are used as explanatory variables (X_{HC}) (see tables 1 and 2) to determine whether a child was employed in the last two months

$$\text{logit}(p_C) := \ln\left(\frac{p_C}{1-p_C}\right) = \alpha + \beta'X_{HC} \tag{3}$$

where p_C = Probability (Ind_IsThisChildLab | X_{HC})

The above econometric approach is to estimate the odds of child labor by using binary multiple logistic regression.

6   Descriptive Statistics

For the households who are working in the carpet industry in Kathmandu Valley this survey estimates a mean household size of 4.8 ([4.6 ; 4.9]$_{95\%}$ $_{CI}$). The mean monthly income is 5,535Rs and the mean per capita income of the household is 1,284Rs ([1,229 ; 1,340] $_{95\% \, CI}$). According to the Nepal Living Standards Survey Report (1996), the per capita income was 2,007 Rs for Kathmandu and 641 Rs for the whole country. The average per capita income in the carpet belt of Kathmandu Valley (1,284 Rs) is significantly lower than that of the overall per capita income estimated in 1996 for the Kathmandu Valley (2,007 Rs); but the households who are working in carpet industries in Kathmandu Valley have a higher per capita income than in the whole country estimated in 1996 (641 Rs). This immense wage gradient between Kathmandu Valley and the rest of the country might induce an intra country migration of child laborers to Kathmandu Valley.

The mean of the household's monthly expenditure is estimated as 4,469Rs. The estimated mean consumption expenditure of the household is 83% ([81 ; 85]$_{95\% \, CI}$ of their income, and the estimated net savings rate is 12% ([11 ; 14]$_{95\% \, CI}$ as the monthly savings amount to 665Rs, and the remaining 4-5% of the income is assumed to be spend to repay a household loan. The net savings per household in this study are derived from the total income of a household from all sources minus the consumption expenditure during the reference period and loan payment. Consumption expenditure includes the amount spent by a household on food and non food items. Almost 71% of the household heads are illiterate and 29% of them have at least primary education. The mean age of the head of the household is 38. The estimated mean unpaid loan of the household is 2,906 Rs ([2045 ; 3767]$_{95\% \, CI}$.

From survey data it is estimated that 91% of the household members joined their first job already in their childhood. The mean age of first joining a profession is 11 (median and mode age is 10). It follows that almost all household members were children when they joined the first job. The mean number of children in the household is 1.81 ([1.70 ; 1.92]$_{95\%}$ $_{CI}$. The mean number of children going to school is 1.05 ([0.96 ; 1.14]$_{95\% \, CI}$. The mean age of starting school is 8 years for children (CI$_{95\%}$: [7 ; 8]). 61% of the total households have at least one child laborer. 55% of the total working children are living in non labeling households[5] but 61% of

---

[5] No household member (adult) is working in the labeling industry.

the total not working children are living in labeling households[6]. On average 53% ([46 ; 60] $_{95\% \text{ CI}}$) of the children are working up to 8 hours and of them 27% ([21 ; 34] $_{95\% \text{ CI}}$) come from labeling households and 26% ([20 ; 32] $_{95\% \text{ CI}}$) come from non labeling households.

Roughly 29% ([23 ; 35] $_{95\% \text{ CI}}$) of the total child laborers work more than 8 hours up to a maximum of 14 hours per day in both labeling and non labeling households. Of them 12% ([7 ; 16] $_{95\% \text{ CI}}$) come from labeling households and 17% ([12 ; 22] $_{95\% \text{ CI}}$) come from non labeling households.

Almost 18% of the child laborers work more than 14 hours per day in both labeling and non labeling households. Of them 6% ([3 ; 10] $_{95\% \text{ CI}}$) work in labeled households and 12% ([8 ; 17] $_{95\% \text{ CI}}$) work in non labeling households.

Hence, exploitation in terms of working hours is higher in the non labeling households than in the labeling households. Figure 2 provides the estimated percentage of different kinds of child work by occupational categories. The survey data estimates that almost 74% of the children are involved in carpet weaving on a part time[7] and/or full time basis[8]. However, less children are working in agriculture (5%) and spinning (5%)[9]. Sometimes, children are involved in a combination of works like spinning and weaving, agriculture and weaving etc. But the occupational categories divided in Figure 2 are based on the main/principal job they perform in the last two months of the survey time.

---

[6] At least one member (adult) is working in a labeling industry and no member is working in the non-labeling industry.

[7] Part time work is defined by the work length of less than 20 days (8 hours per day) in the last two months.

[8] Full time is defined by the work length of more than 20 days (8 hours per day) in the last two months.

[9] Please consider that the survey was focused on the carpet weavers' household in Kathmandu Valley.

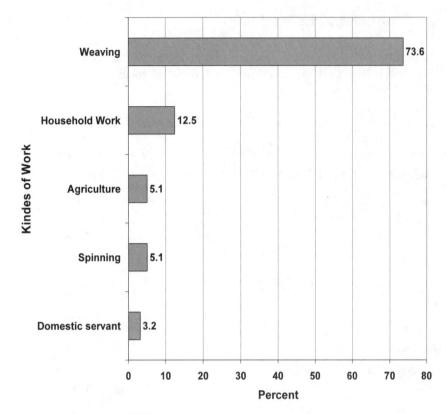

Source: Own Survey (2004)

Fig. 2. Different kinds of work by children

According to this study the incidence of respiratory disease among the carpet workers is nearly 40%. For each episode of illness, respiratory problems are recorded as the most acute diseases. This indicates the susceptibility of these workers to respiratory illness due to exposure to dust from the work, combined with awkward posture, cramped working environment and lack of access to proper health care facilities (KC 2002). Almost 80% of the carpet workers at Kathmandu valley are living inside the factory premises at night. Therefore, the wool dust can cause allergies of the respiratory system and long term exposure may result in obstructive lung diseases.

7   Econometric Estimates

The results of testing the influence of variables on the chance of child labor at the household level (1) or the individual level (2) are shown in Tables 3 and 4 respectively:

Table 3. Logit regression (3.2.1) results for the probability of child labor (household level, N = 410)

Analysis of Maximum Likelihood Estimates			Odds Ratio Estimates	
Parameter		Estimate	Point Estimate	90% Confidence Limits
Intercept		0.79		
HH_IsAnybodyInLBLInd	Registered vs Unregistered	-0.37***	0.48	0.30 0.77
HH_IsAbsDolPov	No vs Yes	0.82	5.10	0.93 28.18
HH_HoH_Sex	Female vs Male	-0.15	0.74	0.30 1.87
HH_HoH_Edu	At least primary education vs No education	-0.39**	0.46	0.27 0.79
HH_IncGT14		-0.78**	0.46	0.26 0.82
HH_N_ChildLE14		1.30***	3.69	2.45 5.54
HH_Debts		0.15*	1.16	1.00 1.33
HH_HoH_Age		0.22**	1.24	1.05 1.46
HH_N_Child0514School		-1.27***	0.28	0.20 0.39
HH_Size		-0.42***	0.66	0.51 0.85

***,**,*: Significant at 1%, 5% and 10% respectively.
Source: Own regression results.

i) The labeling status of a household is an important factor in decreasing child labor participation. A comparison of Tables 3 and 4 shows that for each family as well as for each child, the magnitude of the estimated child labor decreases with labeling NGO intervention. The estimated odds ratio of the labeling status are 0.48[10] for the family-wise regression. This means that the odds of having a child laborer in the family not being assisted by an NGO are more than 2 times[11] the odds of having a working child in an NGO-assisted family. For the child-wise model we get an odds ratio of 0.12 which means, that the odds for a child from an unassisted family to work are more than 8 times[12] higher than the odds for a child to work from an NGO-assisted family. Thus, the null hypothesis "NGO has no influence" in model (5.2.2) and (5.2.3) is not only clearly rejected but also the NGO factor turns out to be the most important factor in preventing child labor.

ii) Following the luxury axiom[13] of Basu and Van (1998), this study tests whether there is a relationship between child labor and adult income ('HH_IncGt14' scaled adult's income in 5,000 Rs). It can be concluded that the sign and the statistical significance of the estimated adult income coefficient support the Basu and Van model. The estimated odds ratio for adult income are 0.46 in the household level regression and 0.64 in the individual level regression. This means, that for each additional 5,000Rs in the family income, the odds for child labor are more than halved (44%) by each 5,000 Rs more (household level) or around 36% (individual level) lower. This shows a strong and negative association between the adult income and child labor in the household.

iii) In the household level and individual level regressions, there is a positive correlation between child employment and family debts ('HH_Debts' scaled household's debt in 5,000 Rs). In both cases, the odds are increased by around 8 to 16%. That means that the odds of child employment are increased by around 8 to 16% if the debt burden of the household rises by each 5,000 Rs.

---

[10] In Table 7.1 the point estimator of the odds ratio of HH_isAnybodyInLBLInd of registered vs. unregistered is 0.481 which is defined as:

$$0.48 = \frac{odds(\text{any child in the family wor king} \mid \text{any one in family in registered industry})}{odds(\text{any child in the family wor king} \mid \text{all in family in unregistered industry})}$$

[11] 2.08 = 1 / 0.48

[12] 8.33 = 1 / 0.12

[13] The family will send the children to the labor market only if the family's income from non child labor sources drops significantly.

Table 4. Logit regression (3.2.2) results for the probability of child labor (individual level, N = 525)

Analysis of Maximum Likelihood Estimates			Odds Ratio Estimates		
Parameter		Esti-mate	Point Esti-mate	90% Confi-dence Limits	
Intercept		0.98			
Ind_NGOAssistChild	Yes vs No	-1.08**	0.12	0.02	0.65
HH_IsAbsDolPov	No vs Yes	0.27	1.70	0.43	6.69
HH_HoH_Sex	Female vs Male	0.01	1.03	0.53	1.98
HH_HoH_Educ	At least primary education vs	-0.28**	0.57	0.38	0.85
	No education				
Ind_Sex	Female vs Male	0.22**	1.55	1.11	2.18
Ind_Mother's_Job	Employed vs Housewife	-0.64**	0.40	0.26	0.62
Ind_Mother's_Job	Expired vs Housewife	0.37	1.10	0.27	4.48
HH_IncGT14		-0.44*	0.64	0.42	0.97
HH_N_ChildLE14		0.28*	1.33	1.03	1.70
HH_Debts		0.08*	1.08	1.00	1.17
HH_HoH_Age		0.09	1.09	0.98	1.21
HH_N_Child0514School		-0.87***	0.41	0.34	0.51
HH_Size		-0.33***	0.72	0.60	0.87

***,**,*: Significant at 1%, 5% and 10% respectively.
Source: Own regression results.

iv) Improvement in the head of the household's education ('HH_HoH_Educ') significantly decreases the probability of a child's employment in the labor market. This is confirmed by the negative and significant estimates in the odds ratio of 'at least primary education' and 'no education' concerning the variable 'head of the household's education' in both, the individual level and household level regressions. The estima-ted odds ratio for 'head of the household's education' are 0.46 in the family-wise regression and 0.57 in the child-wise regression. This means that the odds of child labor are about 54% and 43% lower for those households where the head of the household completed at least primary school compared with those households where the head of the household has no education. This shows a strong and negative association between the education status of the head of the household and child labor.

v) The age of the head of the household ('HH_HoH_Age' Scaled head of the household's age in 5 years of age) shows a significant and positive effect on child labor supply in household level regressions. The use of children as a form of insurance (Pörtner, 2001) also provides some insight into the role of the 'age of the head of the household' in determining child labor. The idea behind this might be that the older the head of the household is, the more aware will he be of his dependency for livelihood in the future. Child laborers could be seen as an 'economic insurance' in old age for the head of the household. Thus, the probability of a child to work is increasing with the age of the household head. The estimated odds ratio for 'age of the head of the household' are 1.24 in the family-wise regression and 1.09 in the child-wise regression, which means that the odds of child labor are 24% and 9% higher for each 5 years increase of the age of the household head. This shows a strong and positive association between the age of the head of the household and child labor.

vi) The sign of the coefficient for the size of a household 'HH_Size' shows that with an increase in the household size, the probability of child labor decreases in both, the individual level and household level regressions. This is contrary to what would have been expected, however, it might be explained by an increased number of adults - and not children - in the household. In fact, the more adults there are in the household, the less likely it is that a child works. The variable 'total number of children' ('HH_N_ChildLE14') shows a statistically significant and positive relation with the occurrence of child labor. This indicates that the higher the number of children in a household, the more likely it is that some children of this family will go to work. The estimated odds ratio for 'total number of children' are 3.69 in the household level regression which means that the likelihood (odds) of a child to work increases by the factor 3.7 for each additional child in the household. This shows a strong and positive association between 'total number of children' in a family and child labor, which is described frequently in the literature (Patrinos 1997).

vii) The estimated odds ratio for 'Ind_Sex' are 1.55 in the child-wise regression. This means that the odds of child labor are about 55% higher for female child compare with male child. This indicates that the higher the number of female children in a household, the more likely it is that child will work.

viii) Mother's employment plays an important role on child labor supply. This study finds that mother's employment significantly decreases the probability of a child's employment in labor market. This is confirmed by the negative and significant estimates in the odds ratio of 'employed' vs 'housewife'. The odds ratio indicates that the likelihood of a child to work decreases by 60% for each child in a household if his/her mother is employed.

ix) This study neither finds a significant influence of absolute poverty ('HH_IsAbsDolPov' household per capita income less than US$ 1per day) nor a significant influence of the 'gender of the head of the household' ('HH_HoH_Sex') on child labor supply of the household. Although the sample size is relatively high to gain a high power this result is likely to have been caused from the fact that 98% of the households report that they live in absolute poverty (less than US$ 1 income). In addition, most people generally underestimate their income if asked in a survey. Also 93% of the households are male-headed. Thus, influences of the 'head of household's gender' or of absolute poverty on child labor supply might still be hard to detect.

8   Conclusion

This study finds that improvement in the child's and household's welfare through the intervention of social labeling NGOs is an effective way of combating child labor at an individual level. One of the main factors which could influence the success of labeling NGOs is 'monitoring frequency'.[14] However, this study does not consider 'monitoring frequency' as an explanatory variable because of the high collinearity with 'HH_IsAnybodyInLBLInd' (Is anybody of the family working in a labeled industry?) and 'Ind_NGOAssistChild' (Is the child being helped by labeling NGO?). In the household level analysis, the most important variable is the number of the children under 14 years of age; a household with more children is much more likely to send a child to work than a household with less children. A combination of policies like labeling NGO's welfare activities, birth control, access to formal credit market,

---

[14] According to the 'RUGMARK BULLETIN' (2003), the frequency of the factory visits varies from once a week to once in two months, depending on the confidence of Rugmark in the factory's commitment and performance with regard to the non use of child labor.

increase of the adult income, and adult education could be suggested from this study to remove a child from the 'work place'.

References

Allison PD (2003) Logistic regression using the SAS system theory and application. SAS Institute, Cary

Basu AK, Chau NH, Grote U (2006) Guaranteed manufactured without child labor: the economics of consumer boycotts, social labeling and trade sanctions. Review of Development Economics, forthcoming.

Basu A, Van H (1998) The economics of child labour. The American Economic Review 88:412-27

Brown DK, Deardorff AV, Stern RM (1999) US trade and other policy options and programs to deter foreign exploitation of child labour. National Bureau of Economic Research Conference, Cambridge

Cartwright K (1999) Child labor in Colombia. In: Grootaert C, Patrinos HA (eds) The policy analysis of child labor: a comparative study. St Martin' s Press, New York

CBS (1997) Nepal living standard survey report 1996 - main findings (volume two). Central Bureau of Statistics, Kathmandu

CBS (1999) Report on the Nepal labour force survey 1988/99. Central Bureau of Statistics, Kathmandu

CBS (1999) Nepal report on the Nepal labour force survey 1998/99. Central Bureau of Statistics, Kathmandu

Charlé S (2001) Rescuing the 'carpet kids' of Nepal, India and Pakistan. Ford Foundation Report, Spring

CUTS (2003) Child labour in south Asia. Consumer Unity & Trust Society, Jaipur

CWIN (1993) Misery behind the looms: child labourers in the carpet factories in Nepal. Child Workers in Nepal Concerned Centre, Kathmandu

Freeman B Richard (1994) A hard-headed look at labour standards, ILO, Geneva

Edmonds EV, Pavcnik N (2004) International trade and child Labor: cross-country evidence. NBER Working Papers 10317, National Bureau of Economic Research, Inc

Edmonds EV, Pavcnik, N (2005) The effect of trade liberalization on child labor. Journal of International Economics 65:401-419

Graner E (1999) Nepalese carpets: an analysis of export oriented production and labour markets. The Economic Journal of Nepal 22:201-217

Grote U, Basu AK, Weinhold D (1998) Child labour and the international policy debate - the education/ child labor trade off and the consequences of trade sanctions. ZEF-Discussion Papers on Development Policy No1, Center for Development Research, Bonn

Hemmer HR, Steger T, Wilhelm R (1996) Child labour in the light of recent economic development trends. International Labour Organization, Geneva

Hilowitz J (1998) Labelling child labour products. International Programme on the Elimination of Child Labour of the International Labour Organization, Geneva

Jafarey S, Lahiri S (2002) Will trade sanctions reduce child labour? The role of credit markets. Journal of Development Economics 68:137-156

Kumar KCB, Govind G, Adhikari S (2002) Child labour in the Nepalese carpet sector: a rapid as*essment. International Programme on the Elimination of Child Labour of the International Labour Organization, Kathmandu

Kish L (2004) Statistical design for research. John Wiley & Sons, New Jersey

Levy PS, Lemeshow S (1999) Sampling of populations - methods and applications. 3rd edn, John Wiley & Sons, New Jersey

Linkenheil K (2003) Introduction of an appropriate labeling system for Nepal's hand knotted carpet industry. Gesellschaft für technische Zusammenarbeit, Kathmandu

Mansfield ER, Helms BP (1982) Detecting multicollinearity. The American Statistician 36:158-160

Maskus K (1997) Core labor standards: trade impacts and implications for international trade policy. World Bank International Trade Division, Washington DC

Melchior A (1996) Child labor and trade policy. In: Grimsrud B, Melchior A (eds) Child labor and international trade policy, Paris

Neumayer E, de Soysa I (2004) Trade openness, foreign direct investment and child labor. Norwegian University of Science and Technology, Trondheim

Patrinos HA, Psacharopoulos G (1997) Family size, schooling and child labor in Peru - an empirical analysis. Journal of Population Economics 10:387 – 405

Ray R (2000) Analysis of child labour in Peru and Pakistan: a comparative study. Journal of Population Economics 13:3-19

Ranjan P (1999) An economic analysis of child labor. Economics Letters 64:99-105

Rosenzweig MR (1982) Educational subsidy, agricultural development, and fertility change. Quarterly Journal of Economics 97:67-88

RUGMARK (2004) RUGMARK Bulletin 2004. Nepal RUGMARK Foundation, Kathmandu

Schultz TP (1997) Demand for children in low income countries. In: Rosenzweig MR, Stark O (eds) Handbook of population and family economics - Vol 1A. Elsevier, Amsterdam, pp 349-430

UNICEF (2003) Assessment of the memorandum of understanding (MOU) regarding placement of child workers in school programmes and elimination of child labour in the Bangladesh garment industry (1995 – 2001). United Nations International Children's Emergency Fund, New York

Wodon Q, Ravallion M (1999) Does child labour displace schooling: evidence on behavioral response to an enrollment subsidy. World Bank, Washington DC

Appendix

Table A1. Number of households in the Kathmandu Valley, Nepal, 2004

District	Households	Percent
Kathmandu	138	33.7
Lalitpur	128	31.2
Bhaktapur	144	35.1
Total	410	100.0

Source: Own survey

Table A2. Places of interview in the Kathmandu Valley, Nepal, 2004

District	Location
Kathmandu	Bauddha
Kathmandu	Bhungmati
Kathmandu	Chabahil
Kathmandu	Chuchepati
Kathmandu	Jorpati
Kathmandu	Kirtipur
Kathmandu	Mahankal
Kathmandu	Swayambhu
Kathmandu	Koteshwor
Kathmandu	Sallaghari
Lalitpur	Bhaisepati
Lalitpur	Ekantakuna
Lalitpur	Nakhkhu
Lalitpur	Sanepa
Lalitpur	Jawalakhel
Lalitpur	Sat Dobato
Bhaktapur	Surya Binayak
Bhaktapur	Sanothimi
Bhaktapur	Jagati
Bhaktapur	Byasi
Bhaktapur	Thimi

Table A3. Labeling status of households

Labeling Status	Households	Percent
Labeling	229	55.9
Non Labeling	181	44.1
Total	410	100.0

Source: Own survey

Table A4. Labeling status of household members

District	Members of Labeling Households	Members of Non Labeling Households	Total Household Members
Kathmandu	307 48.5%	326 51.5%	633 100.0%
Lalitpur	311 51.9%	288 48.1%	599 100.0%
Bhaktapur	489 66.2%	250 33.8%	739 100.0%
Total	1107 56.2%	864 43.8%	1971 100.0%

Source: Own survey

Economic Analysis of Eco-Labeling: The Case of Labeled Organic Rice in Thailand

Maria Cristina DM Carambas

1   Background

Eco-labeling, like the other types of environmental labeling (i.e. mandatory and self-declarations), is the practice of supplying information on the environmental characteristics of a commodity to the general public (Markandya 1997). As a market-based approach to reduce environmental impacts of production, eco-labeling is applied with the assumption that the purchasing behavior of consumers is not just motivated by price, quality, and health standard, but also by environmental or ecological objectives (Deere 1999). Eco-labeling achieves its environmental purpose by influencing change in the purchasing behavior of the consumers in a way that creates incentives for the production of less environmentally harmful products.

Eco-labeling in the agricultural sector, specifically certified organic products, is still gaining ground. The economic and environmental justifycation for eco-labeling can be considered strong enough to promote its adoption in the developing countries. However, there are issues that remain to be resolved. These include the income risks due to uncertainties in productivity, price premium, and the market[1]; the lack of technology or know-how and support services; and the low awareness of this option among producers. These issues may be highly related to how the government is supporting the eco-labeling activities in the country. In the case of the EU, most governments have supported organic farming and eco-labeling via research and development, education, training and extension, market development, and certification, not to mention the financial support for conversion and continued organic production (Padel and Lampkin 1994).

The lack of clear policies in developing countries about organic farming and eco-labeling can be accounted to inadequate information of governments on how eco-labeling fares environmentally, socially, and economically. This study, therefore, aims to contribute to the available body of in-

---

[1] This refers to the stability in supply and demand for these products.

formation on the costs and benefits of eco-labeling or certified organic farming which are deemed useful for more informed policy- and decision-making by governments. This study particularly focused at how production and marketing of eco-labeled products affect the economic standing of the producers. In this regard, the study undertakes to: (i) estimate the costs and benefits of producing labeled organic rice in Thailand relative to its conventional counterpart, (ii) assess the implication of eco-labeling on the profits received and their distribution along the marketing chain, (iii) assess the factors accounting for the difference in marketing margins of eco-labeled and conventional products, and (iv) examine the determinants affecting farmers' decision to adopt eco-labeling.

2   Methods

2.1   Data Sources and Requirements

Since the study aims to learn from the experiences of a developing country, Thailand was selected as one of the study sites. Thailand has pioneering efforts in implementing eco-labeling programs and already established local standards and certification system for organic products. The choice of commodity was based on two considerations: (i) the extent with which eco-labeling has been applied in the commodity's market, and (ii) its importance in the export market. While both considerations were strategic as they ensured availability and easier data collection, the latter has also been a relevant research concern since most - if not all - labeled organic products of the developing countries are being exported.

Both primary and secondary data were used in the analyses. The data collection, particularly on labeled organic rice, entailed field interviews and correspondence through various media (e.g. telephone, e-mails, and letters). The latter collection method was necessary because data on eco-labeled products are not yet systematically collected and published in local and international statistical books. Primary data were collected through a survey of sample farm households and interviews with exporting firms. Structured questionnaires for each type of respondents, i.e. farmers and exporters, were used to elicit the necessary information. Secondary data, i.e. prices, labor wages, etc., were collected from local ministries/agencies and other concerned international agencies, like the ITC/WTO/UNCTAD, FAO, IFOAM, USDA-FAS, and Fair Trade Labeling Organization in Europe, as well as from several special studies on eco-labeling.

The survey was conducted in areas where both labeled organic and conventional rice are mainly grown and produced in Thailand, Surin and

Yasothorn provinces in the northeast region, and Chiang Rai province in the northern region. The two regions accounted for almost 73% of the harvested rice area. The three provinces together accounted for 10% of the rice areas in the two regions, but most of eco-labeled rice for export comes from these provinces. Almost 70% of the estimated total organic rice areas can be accounted to these provinces (Panyakul 2002). The survey was undertaken in 12 villages covered by five districts with high concentration of farmers producing certified and labeled organic rice.

The survey was conducted from February to May 2003 after a pretest was performed to identify the questions in the questionnaire where respondents may encounter problems in answering. Sampling was strati-fied for both the organic and conventional farms according to farm size. Organic farmers were further stratified based on the number of years into organic farming and about 50% of those with at least five years of experience were randomly picked and interviewed. Some replacements had also been resorted to in view of some constraints. Overall, 123 farm households were interviewed in Thailand.

2.2 Analytical Methods

The following analytical methods were applied respectively to address the main objectives of the study: (i) cost-benefit analysis for the assessment of economic and environmental gains from eco-labeling, (ii) commodity chain analysis for assessment of profit distribution, (iii) ordinary least square estimation of marketing margins, and (iv) LOGIT analysis for the determinants of adopting the required organic production approach in eco-labeling.[2]

Assessment of Costs and Benefits

The cost-benefit analysis included an estimation of the financial performance under certified organic commodity production as well as an assessment of its environmental and health implications. For *financial profitability analysis*, this study estimated the net returns for farmer-producers of eco-labeled products, and compares them with those of conventional farmer-producers. In doing so, costs and returns were first evaluated. In general, the difference in revenues per unit of eco-labeled and conventional commodities will depend on the magnitude of the price premium, if

---

[2] Methods of analyses were condensed to fit publication. See Carambas (2005) for detailed explanation and justification.

any. On the other hand, the costs involved in producing eco-labeled products relate to capital costs due to adjustment to new technologies, additional costs of production and processing, increase in labor requirements, additional costs for raw materials, and costs of testing, monitoring, and certification (Grote and Kirchhoff 2001; van Ravenswaay and Blend 1997). These types of costs were estimated through straightforward accounting.

Relative to *environmental concerns*, the long-term costs and benefits of certified organic production were estimated after accounting for the impact of organic production system on soil fertility. The change in soil quality or soil fertility improvement in organic farms was assessed and valued through productivity-change approach. This approach is frequently used in environmental economics to estimate the indirect use value of ecological functions of a natural resource based on its contribution to market activities (Dreschel and Gyiele 1999). In this study, the natural resource is the soil, and its indirect use value is measured by crop productivity or yield. Productivity change is attributed to the farming system used in the production. This means that all the components[3] of the farming system are considered, in general, to affect soil quality. The intertemporal value of the soil can then be determined through the income stream it generates (Grohs 1994). Since commodity outputs are valued at market prices, the value of soil is likewise expressed in terms of market prices.

Considering this analytical framework, the valuation of the environmental benefit (i.e. soil quality/fertility) involves the assessment of the stream of revenues associated with the trend in productivity in a particular production system vis-à-vis the costs of obtaining such a productivity trend. The computation of costs is straightforward and is ba-sed on the previous computation of production cost. Overall, the assess-ment of productivity change in this study involved estimation of yield response model(s) for the conventional and organic production systems, and assessment of net benefits using the internal rate of return (IRR), benefit-cost ratio (BCR) and net-present-value (NPV) measures.

The estimation of a yield response function assesses the influence or significance of production system variables (e.g. inputs, labor, etc.) on the variations in the productivity of farms producing eco-labeled and conventional commodities. The yield response model is a simplistic model for predicting productivity given other alternatives (e.g. bio-dynamic mo-dels)

---

[3] As Lampkin and Padel (1994) noted, organic farming involves restructuring of the whole farming system. It involves modification of agricultural practices like the use of inputs as well as changes in management and labor to replace inputs that are withdrawn after the shift to organic farming.

that have higher predicting capability. The general form of the quadratic function was used to estimate the yield response as follows:

$$Y_t = \alpha_t + \beta_1 x_t + \beta_2 x_t^2 + \beta_3 Z_{it} + \varepsilon \qquad (1)$$

where Y_t represents the yield; x_t, the rate of fertilizer application; Z_{it}, the vector of other variables such as labor and interaction variables (eg. time*x*-production system) which were included in the estimation of combined yield data from farms producing eco-labeled and conventional products; α and β_i, the parameters; and ε_i, the unexplained term. Most models include biophysical factors and soil properties as explanatory variables for the yield response function. In this study, climate and soil-type factors were controlled by utilizing data from the study areas where these factors are, by and large, homogeneous. The difference in soil quality was taken to be a condition resulting from the difference in farming techniques employed in conventional and organic production systems.[4] These also include the fertilizers and labor used which are distinctly different, either in type or in quantity, in the two farming systems. Never-theless, this study recognized that there may be other possible factors/ farming techniques that cannot be considered in or integrated into the model but which may account for significant difference in soil quality in the two production systems, e.g. use of cover crops. In this regard, a dummy variable for production system is included in the model to capture the impact of these farming techniques.

The combined yield response model for conventional and organic production systems that was estimated is as follows:

$$Y = a + b_1 FER + b_2 FER^2 + b_3 LAB + b_4 LAB^2 + b_5 PS + b_6 T + b_7 FERxT + b_8 PSxT + \mu \qquad (2)$$

where Y, FER, LAB, PS, T, FER*x*T, and PS*x*T are yield, fertilizer, labor, production system (binary variable: organic = 1; conventional = 0), time trend index (based on the year of collected data: e.g. 1998=1, 1999=2, 2002=5), interaction of time and fertilizer, and of time and production system. Y is measured by a pooled time series and cross section data.

A quadratic yield response function such as Equation (2) has been widely used for yield response models. The specification of the quadratic function was based on economic theory and agronomic considerations (Larson et al. 2001) where yield-enhancing inputs were allowed to exhibit

---

[4]   It should be emphasized that this assumption was adopted in this study since data measuring annual changes in soil quality attributes were not available. These data would have been also useful in identifying which of the specific soil properties could account for the differences and variations in yield.

diminishing marginal productivity. Inasmuch as the impact of most farming techniques[5] on soil quality are realized in the long-term, the time trend index variable, T, was included to capture the expected long-term yield benefits from the two production systems. Including time trend index in a production function is a standard method for modeling technical change that refers to any kind of shift in the production function. This also reflects the fact that changes in soil or yield may be observable only in the long-term (FAO 1998). Linear interaction terms are used to evaluate potential complementary and competitive technical relationships among relevant variables (Debertin 1986). The interaction terms included in E-quation (2), FERxT and FSxT, have the same intent of determining the long-term impacts of fertilizer and the production system, in general, in terms of yield change. Given the estimated long-term yield from conven-tional and organic farms, the environmental impact of eco-labeling was evaluated by assessing the net market value of the change in yield during a time period.

As for *the health impact*, willingness-to-pay (WTP) is used to measure the economic value of the good or service, i.e. improving health. In this study, information on value, if any, placed by the respondents on the change in environmental amenity that subsequently affect their health, were obtained by directly asking the respondents on how much they value the change, if any. To verify the results, data on the cost of illness, if any, and the cost of averting activities were also asked. A contingent valuation was undertaken in view of the difficulty of getting reliable data on costs of illness and averting activities and information on the causality between the illness and exposure to chemicals and pesticides which are the sources of change in the environmental amenity. The study's reliance on self-reported incidence of disease posed two problems. First, there may be subtle but serious long-term adverse effects to pesticide applicators that they may not be aware of. There might also be health effects to the respondent's families which might affect his utility. These might not be fully captured by the responses. Thus, their self reports could lead to serious underestimations of the health consequences. On the other hand, they could also report illnesses that have nothing to do with the pesticides. This may then lead to a serious overestimation of health effects. In such cases, there would be no assurances that these offsetting influences would cancel out. In addition, there were farmers who, having had no experience of sickness, have expressed willingness to pay for the general reason of having a 'healthier life'.

In view thereof, the contingent valuation undertaken in this study asked the following: (i) concerning the conventional producers, the farmers were

---

[5] Examples are the use of rotation, integrated and/or cover cropping.

asked the minimum and maximum amounts they were willing to pay in terms of a change in productivity in order to attain a healthier farming life by shifting to an eco-friendly production system; and (ii) in the case of the producers of eco-labeled products, they were asked the minimum and maximum amounts they were willing to pay in terms of a reduction in price premium in order to continue a healthier farming life. As Freeman III (2003) noted, willingness to pay can be measured in terms of any other good that mattered to the individual. In this study, these were two `goods´ that were considered relevant: price premium for the organic producers, and yield for the conventional producers. These factors, i.e. the presence of a price premium and the possibility of a yield reduction, summarize the major issues related to organic farming. Instead of money which is the usual form of payment asked in contingent valuation, these goods are con-sidered realistic form of payments from the farmers. They can also be compared in monetary terms as crop yields have market values.

Distribution of Profits

A straightforward assessment of the financial position (in terms of profits received) of the agents or key players in the marketing chain, i.e. from production through processing to export, as market conditions change, was also undertaken. As this study focuses on the production side of the mar-ket, the measure of benefits is in terms of profit or the firms' objective function. In addition, as eco-labeling results in product differentiation or a specialized market in the case of labeled organic products,[6] the analysis of profit distribution along the marketing chain would entail several econo-metric estimations of profit functions for conventional and eco-labeled markets at each level of the marketing chain. This analytical ap-proach is constrained by limitation of time series data on costs and profits at the processing and export market level. At the time of interview, there were only 2 major exporting firms of labeled organic rice. Two conven-tional rice exporters were also interviewed with regard to this study. The assess-ment of profit distribution in this study, therefore, is a static comparison of profits in the markets for the differentiated products. In particular, the changes in the prices received and the costs incurred from production to

---

[6] Eco-labeling, like other forms of labeling, signifies quality and is a basis for product differentiation (Caswell and Mojduska 1996; Roe and Sheldon 2000; Antle 2001). This implies that a primary result of eco-labeling is to create dif-ferentiated markets for labeled and conventional counterpart of a particular commodity. In fact, in the case of labeled organic products, the market is con-sidered a niche (Lohr 2001) or a specialized market.

the exportation of eco-labeled and conventional products are computed at one common time period, and compared at each level of the marketing chain.[7] Since the marketing chain for labeled organic products that are produced in developing countries commonly ends in the export market, the changes in profits and profit distribution were demonstrated at the production and export market levels of labeled organic rice. Assess-ment of profits at the processing level was not possible because for labeled organic rice, this activity is, in most cases, undertaken by the exporters.

The assessment of the distribution of profits considered the general profit function which is defined as follows:

$$\prod (p, w) = \max_x p \bullet f(x) - wx \tag{3}$$

where $\prod$ = profit, p = price of the product, $\mathbf{w}$ = vector of prices corresponding to input vector, f(x) = production function .

In the context of this study, four profit functions are relevant in the profit distribution assessment. These are Π^{FN}, Π^{XN}, Π^{FE}, and Π^{XE}, where Π^{FN} = profit function for the producers of conventional rice, Π^{XN} = profit function for the exporters of conventional rice, Π^{FE} = profit function for the producers of eco-labeled rice, and Π^{XE} = profit function for the exporters of eco-labeled rice. Each of these profit functions faces a different set of output prices and vector of inputs. A total profit identity, Π^{TN} (for conventional rice) and Π^{TE} (for eco-labeled rice), can be derived as:

$$\Pi^{TN} = \Pi^{FN} + \Pi^{XN} \tag{4}$$

$$\Pi^{TE} = \Pi^{FE} + \Pi^{XE} \tag{5}$$

---

[7] This approach is akin to the framework of commodity chain analysis which looks at the financial and economic position of different agents along the length of a production chain. This framework, however, specifically provides a methodological means for analyzing the political economy of global production and trade by Gereffi (1994 and 1999). Although two of the most important dimensions of the analysis are the governance structure and the institutional framework along the chain, the key aspect of the analysis is the location of profits within a chain (Raikes et al. 2000). In this respect, this study's approach can also be seen as an adaptation of the application of commodity chain analysis.

The distribution of total profits between producers and exporters, respectively, can be derived as:

$$\Pi^{FN} / \Pi^{TN} \text{ and } \Pi^{XN}/ \Pi^{TN} \text{ } Conventional\ Rice \qquad (6)$$

$$\Pi^{FE} / \Pi^{TE} \text{ and } \Pi^{XE}/ \Pi^{TE} \text{ } Eco\text{-}labeled\ Rice \qquad (7)$$

Finally, the difference between Equations (6) and (7) provides an assessment of the variation in profit distribution between the two markets.

Determinants of Marketing Margins

An econometric analysis of marketing margins was undertaken to compare conventional and eco-labeled markets in terms of the magnitude and significance of the impacts of demand, supply, and marketing cost in explaining marketing margins. While marketing margins provide neither a measure of farmers' well-being nor of marketing firms' performance, they give an indication of the performance of a particular industry (Tomek and Robinson 1990), or an indication of the market's structure and efficiency.

As Wohlgenant (2001) showed, a marketing margin model could be derived from an analysis of a market equilibrium. Based on a structural model that summarizes the relationships of relevant endogenous and exogenous variables through the specification of supply and demand at the farm and retail market levels, marketing margins can be determined for specific values of exogenous and endogenous variables. With marketing margin equation in the form,

$$M = P_r - P_f (Q_f/Q_r)^8 \qquad (8)$$

and given the demand and supply functions for the farm, retail and marketing input markets, Wohlgenant (2001) showed that partially-reduced form equations yield the following relationships for an econometric estimation of the retail-to-farm price linkage:

$$P_r = P_r (Z, W, T, Q_f) \qquad (9)$$

$$P_f = P_f \ (Z, W, T, Q_f) \qquad (10)$$

$$M = M (Z, W, T, Q_f) \qquad (11)$$

---

[8] It should be noted that the marketing margin is intended to measure the per-product unit costs of assembling, processing, and distributing foods from the farm. Allowing the input-output ratio (Q_f/Q_r) to change represents an efficient utilization of marketing inputs (Reed and Clark, 1998).

where P_r = retail price, P_f = farm price, M = marketing margin, equal to P_r − P_f, Q_f = quantity of the farm input, Q_r = quantity of the retail product, Z = retail demand shifters, W = marketing input prices, and T = other exogenous marketing sector shifters (such as time lag in supply and demand, risk, technological change, quality and seasonality, etc.).

As noted by Wohlgenant and Mullen (1987), since shifts in both demand and supply can cause the output and the retail price to change, a complete analysis of price spread or marketing margin is only possible through an analysis of the complete set of market-behavior equations. However, on the basis of data constraints, a number of analyses on marketing margins used reduced-form models. In Wohlgenant's (2001) empirical model, for instance, Equation (11) was used to estimate mar-keting margins.

In addition to this model, there are four other marketing margin models, based on Wohlgenant (2001) and Lyon and Thompson (1993) which can be used alternatively as reduced-form models. These are:

$$M = f(P_r, W, T) \quad \textit{Mark-up Model} \tag{12}$$

$$M = f(P_r, P_r Q_f, W, T) \quad \textit{Relative Price Spread Model} \tag{13}$$

$$M = f(Q_f, W, T) \quad \textit{Marketing Cost Model} \tag{14}$$

$$M = f(P_f, E_t[P_{ft+1}], W, T) \quad \textit{Rational Expectations Model} \tag{15}$$

where $E_t[P_{ft+1}]$ = expected value of farm price at time t+1.

While the choice of the model(s) had to depend on the significance of the estimation results, the econometric estimations made in this study did not include the mark-up, the marketing cost, and the relative expectations models. This decision was based on theoretical grounds. Lyon and Thompson (1993) had shown that the reduced-form models, particularly Equations (12) to (15), have varying importance in explaining marketing margins depending on spatial and temporal aggregation of data. However, the justification for the specification of the mark-up model is primarily empirical (Wohlgenant and Heidacher 1989). The other models have strong theoretical bases and, thus, render themselves potential alternative marketing margin models. As for the rational expectations model, its assumption on the influence of cost of inventories in price determination is considered irrelevant in the case of eco-labeled commodities given currently low production of these products. In addition, the proposal that the current and past values of farm price affect retail price is also not relevant for the eco-labeled products since prices, both at the farm and consumer levels, are still bilaterally negotiated. Also, rational expectations model is specified for the short term but the data used in this study were on annual basis.

The choice for the reduced-form model derived from the structural model (Equation (11)) and the relative price spread model (Equation (13)) is consistent with the conceptual framework put forth by Gardner (1975). Gardner's framework emphasized the relevance of marketing costs, farm supply, and consumer demand in the determination of price spread. These factors are all represented in the two models. In Equation (11), Z represents the consumer demand factor. In the relative price spread model, the quantity of output and the retail price are the avenues through which the shifts in retail demand and supply are manifested (Wohlgenant and Mullen 1987). The marketing cost model shown is an alternative way of obtaining the relative price spread model will not be estimated, too. As this model is expected to be generally significant given specific data on various marketing inputs and costs, this is unlikely to be the case for eco-labeled markets where official data and statistics are still lacking. In general, the relative price spread model is expected to perform well considering the results of previous studies of Wohlgenant and Mullen (1987), Marsh (1991), Lyon and Thompson (1993), and Richards et al. (1996).

However, it should be pointed out at this point that there might be econometric constraints in estimating this model due to the appearance of an endogenous variable like retail price on the right-hand side of the equation. This issue has rarely been questioned in the literature. In this study, attempts were undertaken to address the issue but nevertheless raises some caveats in the interpretation of the results. There are two ideas to partly address this issue: one is a conceptual clarification and the other is an estimation technique. With respect to the latter, it should be noted that the dependent variable, marketing margin, is not just a difference between retail and farm prices. It also includes the conversion factor in adjusting the quantities. The use of instrumental variable and a two-stage least squares estimation technique may directly but still partly address this issue. In particular, retail price is included in the margin equation as an instrument, i.e. estimating it first using its reduced form in Equation (9). Sargan test was employed to determine whether the instrumental variable used is valid (Gujarati 2003). Based on the results, the econometric estimation of the relative price model may not be reliable in view of the implicit correlation between the dependent variable and one of the independent variables, export price. Though the latter was used as an instrumental variable, results of the Sargan test employed to verify the validity of the instrument cast a doubt that the instrument used is uncorrelated with the error term. In this regard, the general reduced-form model was used in this study to explain the variations in the marketing margins in both eco-labeled and conventional rice markets.

In this study, the analysis of marketing margins which typically refers to the retail-farm price spread, was extended to the export-farm price spread.[9] This seems more appropriate because developing countries are primarily producing labeled organic products for exports. The existing literature has analyzed determinants of marketing margins using structural specifications that involve farm and retail markets. The model specifications used in this study are based on the same structural specifications and derived reduced-form equations. Considering the Law of One Price, or the tendency of prices to equalize across freely trading nations (Houck 1986), the use of the parallel specifications can be justified as the law's assumption of free transfer costs. The fixed exchange ratio of 1:1 implies that when these assumptions are relaxed, the difference in the world price and the domestic price can be explained by exchange rates, transportation costs, and other relevant marketing costs.

Analysis of the Determinants of Farmers' Decision

Finally, the last empirical analysis involved the use of LOGIT regression, i.e. an econometric analysis involving a dependent variable that signals a probability condition for adopting organic farming. As Gyawali et al. (2003) noted, economic theory provides limited guidance in the selection of variables to explain the participation behavior of farmers. However, the findings of previous studies provided relevant inferences on the factors to be considered in this study. In this regard, the specific hypotheses with respect to the direction of the effects of each factor are based on the general findings of previous researches on similar topics.

It is hypothesized in this Chapter that the farmers' decision to adopt or not to adopt organic farming is influenced by a wide range of factors that can be categorized as: (i) socio-economic characteristics of the farmers, (ii) characteristics of the farm, (iii) factors relating to farmers' support/ assistance, (iv) farmers' perceptions on the impacts of adopting the farming approach required for eco-labeling, and (v) other economic factors.

The *socio-economic characteristics* included in this analysis are sex, age, civil status, education, farming experience, major source of income, and the level of income. During the survey in Thailand, a number of farmers mentioned the importance of their own children's health in their deci-

---

[9]   Ahmed and Rustagi (1987) also considered in their analyses of marketing margins the export prices for foreign consumers as the other end of the market levels analyzed.

sion, at least, in converting to organic farming. In view thereof, the number of children and the number of family members working in the farm were also included in the model.

Among the *farm characteristics* considered relevant in the analysis are farm size, number of plots, and ownership status. *Farmers' support/ assistance* (informational and technical) has also been cited in the literature as significant factors to persuade farmers to adopt a certain technology or production approach. Farmers' awareness, availability of information, and farmer organization's /association's role in promoting the adoption are included in this category.

As Burton et al. (1998) noted, farmers' opinions and attitudes account for the probability to adopt a new production approach. In this study, the *perception of the farmers* concerning the impact of certified organic farming on health, environment, yield, cost, and income are expected to be relevant indicators of the farmers' decision to adopt. Lastly, the *other factors* relate to the revenue/income effect of adopting eco-labeling and to the actual experience of the negative health effects of the conventional production system. For the purpose of this study, only the experience of sickness was included under this category because the farmers' decision to adopt or not is likely to be based on their perception on revenue/income effect of organic farming. As income effect is actually felt only after adoption, actual revenues would influence the farmers' decision to continue organic farming or not.

3   Results

The following are the major findings of the study:

3.1   Assessment of Costs and Benefits

On Production Costs and Returns

Based on the average mean of the survey data, rice farms producing and marketing labeled organic rice performs at par with the conventional counterpart (Figure 1). Despite increased cost due to higher input, labor and certification costs, net revenues per metric ton (MT) unit of organic rice are higher than the conventional rice because of price premium.[10] On the

---

[10] Price premium is expressed as percentages by which the prices of labeled organic products are above the prices of similar conventional products.

average, price premium for labeled organic is 100% for farmers selling to the major NGO exporter and 4% for those selling to private companies.

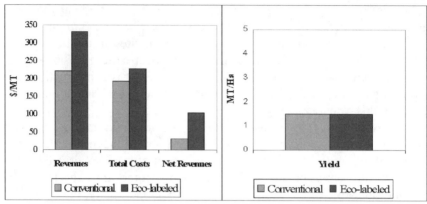

Source: Own computations/illustrations based on survey data.

Fig. 1. Yield and revenue comparisons for labeled and conventional rice in Thailand (in milled paddy rice), 2002

On Productivity Assessment

The yield response function[11] was estimated (Table 1) using a panel data of farmers' yield during the last five years, and was applied in obtaining a deterministic prediction of organic and conventional rice yield growth in the long term.[12] Using the yield forecasts and assumptions[13] on other relevant variables, the long-term benefits obtained due to productivity change in the two production systems were computed and assessed using benefit-cost ratio (BCR) and net present values (NPV).

---

[11] The relevant final model is a Cobb-Douglas function rather than a quadratic function which had low coefficient of determination (R-squared).

[12] The prediction holds labor and fertilizer inputs constant, and relies on the technical change and long-term impact of the production system.

[13] Starting yields and farm prices for conventional and organic rice are based on 2002 official national data and the exporting companies, respectively. Input costs are based on the inflation rate and average growth in wage during the last 10 years. Exchange rate used is fixed at the 2002 level, while the discount rate used is 15%.

Table 1. Estimated rice yield response functions for conventional and organic production systems (double log functional form)

Variable	Pooled	Organic	Conventional
1. Constant		5.11	4.06
		(25.83)***	(11.27)***
2. Production System			
a) Conventional	4.05		
	(11.47)***		
b) Organic	4.91		
	(2.53)***		
3. Fertilizer			
a) Conventional	0.35		0.33
	(3.61)***		(3.49)***
b) Organic	0.05	0.05	
	(1.89)*	(1.88)*	
4. Labor	0.26	0.16	0.28
	(6.18)***	(1.73)*	(5.41)***
5. Time			
a) Conventional	0.01		0.01
	(1.05)		(1.14)
b) Organic	0.03	0.03	
	(2.48)**	(2.35)**	
Observations	336	126	210
	0.92	0.92	0.92

Source: Own computations
Notes: |z| statistics in parentheses
Significance levels: * (α=0.10), ** (α=0.05), ***(α=0.01).
a) The production system is included as a dummy variable. In this model, the conventional production system appeared as a constant.
b) The original coefficient of the dummy was 0.86. For presentation purposes, this value is already evaluated with respect to the constant or the second category of the dummy on the production system, i.e. 4.04+0.86.

Results show that, in general, the economic value of the environmental benefits of organic farming, even with price premium for those that are marketed as labeled organic rice could only be realized in the long-term. If there are no price premia, the net present value of the benefits for producing eco-labeled commodities will be lower than that of the conventional counterpart. In particular, Table 2 shows that it takes about 15 years for the organic production system to reach the same productivity level as that of the conventional production system, i.e. 2.24 MT/ha.[14] The realized long-term NPV of producing eco-labeled rice through organic farming is positive, and greater than the NPV obtained from the production of conventional rice if the price premium for eco-labeled rice is maintained (Scenario 1). Without price premium (Scenario 2), both BCR and NPV are lower for eco-labeled rice compared with those of the conventional rice. During this 15-year period, therefore, producer of eco-labeled rice may still find productivity change benefits to be negative, i.e. in terms of net economic value of the difference in crop yield between the two production systems.[15]

Over a longer time period, rice yield from organic production surpasses its conventional counterpart, as shown in the 30-Year time horizon in Table 2. BCR and NPV have also significantly improved. However, although the productivity benefits have improved, the magnitude is rather small particularly in terms of NPV. Considering the declining discount factor over time,[16] a positive productivity benefit may come only in a very long-term period. Notwithstanding this, it should be noted that with the presence of price premium for labeled organic rice, economic benefits received by producers of eco-labeled products are positive and higher than those received by conventional producers, as seen in Scenario 1. In general, the results of the productivity change analysis have been found consistent with general knowledge on organic farming or any soil-conserving practices, i.e. yields are expected to differ more significantly only in the future (Enters 2000). However, as a caveat for interpreting the results, it should be noted that the models face several limitations. For example, due to lack of data, the variables included in the model may have limited the explanatory power of the model. In addition, the analysis does not take into account the

---

[14] This is consistent with the findings of Lampkin and Padel (1994) that absolute yield levels under organic management are increasing over time but at a slower rate relative to the conventional system.

[15] This result is consistent with a conclusion of Lampkin and Padel (1994) that based on the farm-level studies, price premia are needed to compensate for reduced output and increased labor use. They noted from the studies by Braun (1994) and Zerger and Bossel (1994) that farm incomes fall when farmers do not obtain premium prices. This is, however, assuming that all farms have converted to organic farming. Meanwhile, other studies indicated that incomes could be higher as a result of reduced output supply. Nevertheless, Lampkin and Padel warned that this last result should consider the problem in extrapolating from the current state of organic farming which is only a small part of agricultural activities.

[16] This implies that although total NPV is higher in a 30-year period than in a 15-year estimation period, the annual rate of change declines every year.

possible impact of production of eco-labeled products on prices. Considering the small production size of these products, we posit that its impact on price may be neglected. In view of this, the estimates should be interpreted together with some results on the sensitivity of the estimates on deviations in costs and benefits.

Table 2. Cost-benefit analysis of producing labeled organic and conventional Thailand rice: 15-year and 30-year time horizon

	End Period Yield	Mean Yield	Benefit-Cost Ratio	Net Benefits	Net Present Values (NPV)
	(MT/ha)	(MT/ha)	(BCR)	(US$)	(US$)
15-Year Time Horizon					
Scenario1: Eco-labeled Rice Receives Price Premium					
(i) Conventional Rice	2.24	2.06	2.58	4,043.97	1,350.00
(ii) Eco-labeled Rice	2.25	1.79	2.69	5,994.05	1,799.90
(iii) Productivity and Price Benefits				1,950.08	449.90
Scenario 2: Eco-labeled Rice Does Not Receive Price Premium					
(i) Conventional Rice	2.24	2.06	2.58	4,043.97	1,350.00
(ii) Eco-labeled Rice	2.25	1.79	1.64	2,276.47	612.33
(iii) Productivity Benefits				-1,767.50	-737.67
30-Year Time Horizon					
Scenario1: Eco-labeled Rice Receives Price Premium					
(i) Conventional Rice	2.68	2.26	3.13	14,159.55	1,767.22
(ii) Eco-labeled Rice	3.76	2.39	4.07	28,109.82	2,605.33
(iii) Productivity and Price Benefits				13,950.27	838.11
Scenario 2: Eco-labeled Rice Does Not Receive Price Premium					
(i) Conventional Rice	2.68	2.26	3.13	14,159.55	1,767.22
(ii) Eco-labeled Rice	3.76	2.39	2.54	14,119.89	1,058.16
(iii) Productivity Benefits				-39.66	-709.06

Source: Own computations
[a] Computed as the difference of benefits, i.e.Net Benefits and NPV, received by producers of labeled organic and conventional rice, that is, (ii) less (i).
[b] Productivity benefits also refer to difference in benefits, i.e. Net Benefits and NPV, received by producers of eco-labeled and conventional rice.

On Health-Related Effects of Organic Farming/Eco-labeling

The economic value of health benefits of organic farming are revealed by the willingness to pay (WTP) of both the conventional and organic farmers. WTP is measured in terms of yield (for conventional producers) and price premium (for organic producers) that farmers are willing to forego in order to practice organic farming and reap the health benefits associated with it. Although the amount that the conventional rice farmers are willing to pay is less than the amount that the producers of organic rice are willing to pay, the former is nonetheless willing to pay about 21% of its expected total production revenue. Organic farmers can pay up to 49% of their potential revenues with price premium accorded to labeled organic products. Potential revenues were computed based on the potential yield of each group and the farm price in 2002.

3.2   Profit Distribution Analysis

Table 3. Relative profits at the marketing chain[a]

| Thailand Rice | Relative Profit [b] (Eco-labeled vs Conventional) | |
	Farm Level $(\Pi^{FE} / \Pi^{FN})^c$	Export Level (Π^{XE} / Π^{XN})
Product channeled through:		
i. NGO	17.97	17.96
ii. Private exporting company	3.73	19.45

Source: Own computations.
[a] Profits are in per-unit basis for comparison purposes; calculated using 2002 data.
[b] Ratio of profits received either by farmers or exporters at the eco-labeled markets and profits received at the conventional market.
[c] Π^{FN} = profit for the producers of conventional rice; Π^{XN} = profit for the exporters of conventional rice; Π^{FE} = profit for the producers of eco-labeled rice; and Π^{XE} = profit for the exporters of eco-labeled rice.

The computed relative profits (Table 3) indicate that profits, both at the farm and export levels of eco-labeled rice market, are generally higher than profits at the conventional counterparts. In particular, the findings show that the per-unit profits for producing eco-labeled rice range from about 4 to 18 times that of the profits for conventional rice. Relative profits at the export level are higher than at the farm level. It is noted that rice farmers who are producing for an NGO-exporter in Thailand have higher relative profit than those producing for a private company.

In terms of profit distribution (Table 4), the results show a reduction in the share of profits at the producers' level, i.e. the share of the farmers producing labeled organic rice is lower by about 30% compared to the share of the conventional counterpart. For labeled organic rice produced for private company, the share is lower by about 80%. On the other hand, the share of exporters of labeled organic rice in the profits is higher by about 10 to 30% compared to the share of the exporters of conventional products.

Table 4. Distribution of profits between farmers and exporters, 2002

	Profit Shares (%)		Change in the Profit Share (%)	
	Farm level	Export Level	Farm level	Export Level
Thailand Rice	(Π^{FE}/Π^{TE}) or $(\Pi^{FN}/\Pi^{TN})^a$	(Π^{XE}/Π^{TE}) or (Π^{XN}/Π^{TN})	$[(\Pi^{FE}/\Pi^{TE})$ - $(\Pi^{FN}/\Pi^{TN})]$	$[(\Pi^{XE}/\Pi^{TE})$ - $(\Pi^{XN}/\Pi^{TN})]$
a. Eco-labeled				
Channeled through:				
i. NGO	20.2	79.8	-26.3	9.9
ii. Private company	5.1	94.9	-81.4	30.6
b. Conventional	27.3	72.7		

Source: Own computations.
[a] $\Pi^{TN} = \Pi^{FN} + \Pi^{XN}$ and $\Pi^{TE} = \Pi^{FE} + \Pi^{XE}$.

3.3 Analysis of Marketing Margins

Based on both the trend of prices vis-à-vis marketing margins and the results of the regression (Table 5), it is noted that there is an inconsistency in price determination particularly at the farm level in Thailand. This is evident in the increasing export prices juxtaposed with declining farm prices, and a very low transmission elasticity, i.e. from export price to farm price (Table 5).

In Table 5, it was noted that the marketing margins in labeled organic rice are relatively more elastic with respect to consumer income than for the conventional rice. This shows that it is more difficult for the supply for labeled organic rice to respond to change in demand given the requirements for producing this commodity. Under the assumption of positively-sloped marketing inputs supply, substitution of marketing inputs for farm inputs may be undertaken to increase supply of the final product. However, this could raise consumer prices thereby resulting in higher marketing margins. The empirical observation that demand has greater influence on marketing margins than farm supply was confirmed by result of the regression where the latter's coefficients are generally low in magnitude and in statistical significance. It is interesting to note that organic rice farm supply has positive influence on marketing margins which is contrary to theoretical expectations.

Table 5. Regressions on marketing margins using the general reduced-form model for labeled organic and conventional rice

	Consumer Income (GDP/capita)	Farm Input Supply	Marketing Cost Index (includes fuel and wage costs only)	Constant	Adj. R-squared
Conventional Rice	0.62	-0.07	0.63	-0.13	0.37
	(1.84)*	(0.08)	(2.20)**	(0.64)	
Labeled Organic Rice	1.53	0.53	0.31	-6.03	0.92
	(3.26)**	(4.71)***	(2.49)*	(3.95)*	

Source: Own computations.
Notes: Significance levels: * (α=0.10), ** (α=0.05), ***(α=0.01).

Table 6. Elasticities[a] of farm price with respect to FOB price

	Conventional	Eco-Labeled
	0.94***	0.86
	(0.96)	(0.84)
	[2.37]	[1.80]

Source: Own computations.

Notes: Significance levels: * (α=0.10), ** (α=0.05), ***(α=0.01).
[a] The elasticities of price transmission are obtained by estimating a long-run backward price transmission model, i.e. $\ln(FARMPRICE) = a + b \cdot \ln(FOBPRICE) + c \cdot TIME$.
[b] Values in parenthesis are the adjusted coefficients of determination (Adj. R-squared); Durbin-Watson statistics, in brackets.

Marketing costs were also found to be significant explanatory variable for the variation in marketing margins in conventional rice but relatively less significant for the variation in marketing margins in eco-labeled rice. In this regard, it may be argued that in the case of labeled organic rice, wages and fuel costs may be poor proxies for costs of marketing or that the high marketing margins for labeled organic rice is explained by other factors than marketing costs which is conventionally the primary factor considered in explaining marketing margins. For instance, market power in terms of price determination at the farm level may be looked into.

3.4   Analysis of the Determinants of Farmers' Decision

The significant factors affecting Thai farmer's decision to adopt organic production represent each of the five major categories of determinants discussed in the methods section that are hypothesized to be relevant in explaining the decision of the farmers to adopt (Table 7). Of these factors, socio-economic characteristics, i.e., sex and family size, and farm characteristic, i.e. tenure, have relatively smaller influence on the decision based on the values of marginal probabilities. On the other hand, the marginal probability that a farmer will adopt organic production system increases by an average of 50% when: (a) it is easy to get technical information about eco-labeling, (b) farmers perceive positive yield and environmental effects of organic farming, and (c) farmers experienced sickness in conventional farming.

Table 7. Maximum likelihood estimates and goodness-of-fit measures

Predictors	Coeffi-cient	Odds Ratio	Wald Stat	Change in Probability
Sex	0.94	2.57	1.76*	0.22
Family Size	-0.53	0.59	-2.11**	-0.12
Major Source of Income (Agri=1, Non-Agri=0)	-1.36	0.26	-1.50	-0.26
No. of Plots	0.53	1.69	1.82*	0.12
Land Owner (Yes=1, No=0)	1.55	4.73	1.02	0.37
Access to Technical Information_2 [a]	-0.04	0.96	-0.07	-0.01
Access to Technical Information_3 [a]	3.68	39.73	2.80***	0.47
Expected Yield Impact of Organic Farming_2 [b]	-2.23	0.11	-2.67***	-0.50
Expected Yield Impact of Organic Farming_3 [b]	-2.09	0.12	-2.58***	-0.46
Expected Impact of Organic Farming in Reducing Negative Environmental Impact of Farming_3 [b]	2.48	11.90	1.64*	0.52
Experience of Sickness During Conventional Farming (Yes=1, No=0)	2.60	13.45	3.80***	0.47
Number of observations	118			
LR chi^2 (Prob > chi^2)	58.16	(0.00)		
Pseudo R^2	0.36			

Source: Own computations.
Note: Significance levels: * (α=0.10), ** (α=0.05),
***(α=0.01).
[a] (Difficult=1, Not so Difficult=2, Easy=3)
(1=None, 2=Reducing, 3=Increasing, 4=Cant assess)

4   Conclusions and Policy Implications

This study has shown that financial, environmental and health benefits could accrue to producers of labeled organic rice. However, financial benefits largely depend on the presence of price premium. In this regard, the decision on which farming system would be preferable on the farmers' point of view may have to depend on the extent to which the environmental and health impacts could compensate for the uncertainty of relying on a price premium for financial profits. On the part of the government, it may have to balance the need to address current food security vis-à-vis long-term food security attained through a more environment-friendly use of the soil for crop production. Lampkin and Padel (1994) noted that despite a yield-reducing effect of organic production, organic food can still meet domestic food demands in most countries in the EU. However, the issue of lost opportunity to produce for the rest of the world becomes imperative. In the developing countries, the discourse on the extent to which they should be encouraged will have to balance the importance of quantity, price, and income effects to producers with the price and quantity effects to consumers.

The issue on the physical productivity of the (organic) farming system still remains. Lampkin and Padel (1994) previously stated that yield penalties for producing organic products may have been frequently overstated due to inappropriateness of the comparative-static approach or the analytical model used in assessing productivity impacts. Apart from this, however, it is fundamental to consider that the organic farming system lacks the needed research and technological development. While the model used in this analysis may run the risk of either underestimating or overestimating the productivity potential of the organic farming system, there may be quite an adequate evidence that organic farming system has untapped productivity potential due to lack of government support. The fact that the yield potential had increased from the 1950s to the 1990s (Lampkin and Padel 1994) implies that it may be increased further if adequate governmental support is given. Given this, the current lack of support may have rendered the comparison of its productivity potential grossly misleading. Notwithstanding these issues, the fact that there are farmers who are undertaking and willing to participate in eco-labeling should prompt the government to provide the necessary support to enable them to successfully shift to the desired production system. Overall, the assessment does not and, as intended, should not give exact indications on whether to implement eco-labeling or encourage the adoption of organic farming system. Indeed, at this point in time when the technology for organic farming is still underdeveloped and by itself would not be able to meet the current food demand, the preference on which farming system to undertake should be left to farmers' discretion. However, being free to decide on which to undertake means that the farmers are also provided with not only adequate information but also the necessary technology and support services that will be needed should they decide to undertake an alternative to conventional farming system.

This study has shown the potentially positive impacts of producing eco-labeled products using the organic farming system in the developing countries despite the low level of support that governments provide to the industry. Given this, the provision of governments' support and assistance through, first and foremost, a clear policy towards the promotion of eco-labeling or organic farming could further strengthen the potential of this industry to provide positive and significant economic and environmental impacts. In this regard, there will be a need for credible and effective institutions to implement the policy as well as corresponding support services particularly on research, technology promotion, and extension. As shown in the analysis of determinants of farmers' adoption of organic farming, information on the technicalities of organic farming as well as on the possible impacts of organic farming serves as an important incentive for farmers' decision to adopt this farming approach.

In this study's assessment of profit distribution, it is shown that although profits at the farm level could be higher in absolute terms than those of the conventional counterparts, the latter gets a higher percentage share of the total profits in the marketing chain than the former. While this does not alter the fact that the farmers are better off in participating in eco-labeling, the assessment of profit distribution is relevant as it provides an indication on how the labeled organic product market currently operates. It also raises an issue on whether there are possible measures that can be undertaken to ensure that farmers get optimal economic benefits, without prejudice to the share of the other market participants.

Finally, the analysis of marketing margins shows that high marketing margins for labeled organic rice are not highly explained by marketing costs. Based on the historical trends in prices and marketing margins, there are some indications that marketing margins may also be explained by pricing arrangements between farmers and exporters. Although the study has shown that producer of labeled organic rice have indeed received economic gains through higher prices and income compared to their conventional counterparts, an assessment of price determination in the marketing chain should be undertaken to ensure that all market players have equal opportunities to capture optimal price benefits offered by the market. This would also ensure that these market players are given the right incentives to participate in the production of environment-friendly products.

References

Ahmed R, Rustagi N (1987) Marketing and price incentives in African and Asian Countries: a comparison. A World Bank symposium. In: Elz D (ed) Agricultural marketing strategy and pricing policy, World Bank, Washington DC, pp 104-118

Antle JM (2001) Economic analysis of food safety. In: Gardner BL, Rausser GC (eds) Handbook of agricultural economics - Vol 1B. Elsevier, Amsterdam, pp 1083-1136

Braun J (1994) Impacts of widespread conversion to organic agriculture in the state of Baden-Württemberg, Germany. In: Lampkin N, Padel S (eds) The economics of organic agriculture: an international perspective. CAB International, Wallingford, pp 329-342

Burton M, Rigby D, Young T (1998) Adoption of organic agriculture in Europe: economic and non-economic determinants. Paper based on the research undertaken as part of the Economic and Social Research Council (ESRC) Global Environment Change Project, http://crcnetvividglobalnetau/newsletter/ SeaNews/seaos2htm Accessed in September 2004

Carambas M (2005) Economic analysis of eco-labeling in the agricultural sector of Thailand and the Philippines. Göttingen: Cuvillier Verlag

Caswell J, Mojduszka EM (1996) Using informational labeling to influence the market for quality in food products. American Journal of Agricultural Economics 78:1248-1253

Debertin DL (1986) Agricultural production economics. Macmillan Publishing, New York

Deere C (1999) Eco-labeling and sustainable fisheries. International Union for Conservation of Nature and Natural Resources/ Food and Agriculture Organization of the United Nations, Washington DC/ Rome

Dreschel P, Gyiele LA (1999) The economic assessment of soil nutrient depletion: analytical issues for framework development issues in sustainable land management. International Board for Soil Research and Management, Bangkok

Enters T (2000) Methods for the economic assessment of the on- and off-site impacts of soil erosion issues in sustainable land management 2. International Board for Soil Research and Management, Bangkok

FAO (1998) Evaluating the potential contribution of organic agriculture to sustainability goals. Food and Agriculture Organization of the United Nations technical paper, contribution to International Federation of Organic Movement's Scientific Conference, Mar de Plata, Argentina, http://wwwfaoorg/DO CREP/003/AC116E/ac116e00htm Accessed in June 2004

Freeman III AM (2003) The measurement of environmental and resource values: theory and methods. 2nd edn RFF Press, Washington DC

Gardner BL (1975) The farm-retail price spread in a competitive food industry. American Journal of Agricultural Economics 85:235-242

Gereffi G (1994) The organization of buyer-driven global commodity chains: how the US retailers shape overseas production networks. In: Gereffi G, Korzeniewicz M (eds) Commodity chains and global capitalism, Praeger Publishers, Westport

Gereffi G (1999) A commodity chains framework for analyzing global industries. http://wwwidsacuk/ids/global/pdfs/gereffipdf. Accessed in March 2004

Grohs F (1994) Economics of soil degradation, erosion and conservation: a case study of Zimbabwe. Wissenschaftsverlag Vauk, Kiel

Grote U, Kirchhoff S (2001) Environmental and food safety standards: issues and options. ZEF Discussion Papers on Development Policy 39/2001 Bonn: Center for Development Research (ZEF)

Gujarati DN (2003) Basic econometrics. 4th edn. McGraw-Hill, New York

Gunes I (2004) The change in productivity technique. http://idaricuedutr/igunes/cevre/valintrohtm Accessed in March 2004

Gyawali BR, Onianwa O, Wheelock G, Fraser R (2003) Determinants of participation behavior of limited resource farmers in conservation reserve program in Alabama. Paper presented at the Southern Agricultural Economics Annual Meeting Alabama, USA: Mobile Convention Center

Houck JP (1986) Elements of agricultural trade policies. Macmillan Publishing, New York

ITC/CTA/FAO (2001) World markets for organic fruits and vegetables: opportunities for developing countries in the production and export of organic horticultural products. International Trade Center /Technical Center for Agricultural and Rural Cooperation /Food and Agriculture Organization of the United Nations, Rome

Kortbech-Olesen R (1998) Market prospects for organic food and beverages. Paper presented during the 12th session of the Committee on Commodity Problems Intergovernmental Group on Citrus Fruit Geneva, Switzerland: International Trade Center/ United Nations Conference on Trade and Development/ World Trade Organization

Lampkin NH, Padel S (eds) (1994) The economics of organic farming - an international perspective. CAB International, Wallingford

Larson JA, Roberts RK, Jaenicke EC, Tyler, DD (2001) Profit-maximizing nitrogen fertilization rates for alternative tillage and winter cover systems. Journal of Cotton Science 5:156-168

Lohr L (2001) Factors affecting international demand and trade in organic food products. In: Regmi A (ed) Changing structure of global food consumption and trade. Economic Research Department - United States Department of Agriculture, Washington DC, pp 67-79

Markandya A (1997) Eco-labelling: an introduction and a review. In: Vossenaar R, Zarrilli S, Jha V (eds) Eco-labeling and international trade. Macmillan Press, London, pp 1-19

Marsh JM (1991) Derived demand elasticities: marketing margins methods vs inverse demand model for choice beef. Western Journal of Agricultural Economics 16:382-391

Mäder P, Pfiffner L, Fliessbach A, von Lützow M, Munch JC (1996) Soil ecology: the impact of organic and conventional agriculture on soil biota and its significance for soil fertility. In: Oestergaard TV (ed) Fundamentals of organic agriculture - proceedings of the 11th international scientific conference of the international organisation for organic agricultural movements, 'Down to Earth and Further Afield', Vol II. International Organisation for Organic Agricultural Movements, Tholey-Theley, pp 24-46

Padel S, Lampkin N (1994) Farm-level performance of organic farming systems: an overview. In: Lampkin N, Padel S (eds) The economics of organic agriculture: an international perspective. CAB International, Wallingford, pp 201-219

Panyakul V (2002) National study: Thailand. In: ESCAP (ed) Organic agriculture and rural poverty alleviation: potential and best practices in Asia. United Nations Economic and Social Commission for Asia and the Pacific, Bangkok, pp 173-203

Raikes PM, Jensen MF, Ponte S (2000) Global commodity chain analysis and the French filiere approach: a comparison and critique. CDR Working Paper Series 003/2000 Copenhagen, Denmark: Center for Development Research

van Ravenswaay E, Blend J (1997) Using eco-labeling to encourage adoption of innovative environmental technologies in agriculture. Department of Agricultural Economics Staff Paper 19/1997 East Lansing, USA: Michigan State University

Richards TJ, Ispelen PV, Kagan A (1996) Forecasting retail-farm margins for fresh tomatoes. Arizona State University East: NFAPP 01/1996, http://www.easasuedu /~nfapp/discus/marg796doc) Accessed in March 2004

Roe B, Sheldon I (2000) The impacts of labeling on trade in goods that may be vertically differentiated according to quality. In: Bohman M, Caswell JA, Krissoff B (eds) Global food trade and consumer demand for quality (chapter 10). Kluwer Academic/Plenum Publishers, New York

Scialabba NE, Hattam C (2002) Organic agriculture, environment and food security. http://wwwfaoorg/documents/show_cdrasp?url_file=/DOCREP/005/Y4137E/Y4137E00HTM Accessed in March 2003

Tomek WG, Robinson KL (1990) Agricultural product prices. 3rd edn. Cornell University Press, New York

Wohlgenant MK (2001) Marketing margins: empirical analysis. In: Gardner B, Rausser G (eds) Handbook of agricultural economics - Vol 1B. Elsevier Science BV, Amsterdam, pp 933-970

Wohlgenant MK, Mullen JD (1987) Modelling the farm-retail price spread for beef. Western Journal of Agricultural Economics 12:119-125

Wohlgenant MK, Haidacher RC (1989) Retail to farm linkage for a complete demand system of food commodities. USDA Technical Bulletin 1775, United States Department of Agriculture, Washington DC

Zerger U, Bossel H (1994) Comparative analysis of future development paths for agricultural production systems in Germany. In: Lampkin N, Padel S (eds) The economics of organic agriculture: an international perspective. CAB International, Wallingford, pp 317-328

Eco-labeling and Strategic Rivalry in Export Markets

Arnab K. Basu, Nancy H. Chau and Ulrike Grote

1 Introduction

There are two linkages that constitute the interface between international trade and the environment. The first arises when purely trade driven incentives, rather than environmental considerations, guide production decisions in such a way that environmental exploitation in the name of trade is threatened. These result in scale, composition and technique choices that fail to internalize consumers' preferences with respect to production and process methods (PPMs), or society's preferences with respect to local and transnational environmental commons (Grossman and Krueger 1995, Copeland and Taylor 1994, 1995). The second stems from the possibility that international trade unleashes competitive pressure that put emphasis on policies and technology choices that facilitate cost-cutting (Frankel and Rose 2002, Porter and van der Linde 1995). Here, the concern is over a potential race to the bottom in environmental performance standards, in which trade ties between countries and a vicious cycle of environmental policy *interdependence* are inextricably linked.

In this context, eco-labeling - the provision of information about the environmental externalities associated with the production and consumption processes - holds the promise of cutting through both of these knots. By re-establishing the link between marginal environmental gains and revenue incentives, eco-labeling offers to provide market-based rewards to producers that practice green production methods through a green premium. Concurrently, by rendering the adoption of green technology to a profitable enterprise, incentives to participate in the race to cut costs may be moderated by competition that is based jointly on comparative cost and reputational advantage, backed by the credibility of an environmental performance guarantee.

It is thus perhaps not all that surprising that the adoption of eco-labels in both industrial and agricultural sectors has grown worldwide (Basu, Chau and Grote 2003). Labeling initiatives in agriculture, for example, are particularly notable for their relatively early start. In countries such as Germany, France and Italy, food industry eco-labeling initiatives have been in existence as early as the 1920's. In addition, since global agricultural trade impacts the interests of developed and developing countries, another question that arises is whether eco-labeling can exacerbate income disparities between developed and developing countries, when the latter may be at a disadvantage based both on cost and revenue grounds. In terms of costs, the effectiveness of eco-labeling depends on whether green technologies are readily accessible, and accordingly whether the costs involved in implementing labeling programs can be afforded or even justified (UNDP 1999). In terms of revenue, the relative credibility of labeling programs in developed and developing countries - whether perceived or realistic - may impact terms of trade facing developing country exports in ways that are similar to other non-tariff import barriers (UNDP 1999, Basu and Chau 2001).

In assessing the promise of eco-labeling, therefore, a number of pertinent questions arise. First, do producers behave as though a green premium indeed exists? Second, do strategic interactions between trading partners in their decision to adopt labeling prevail, and if so, has there been a race to the bottom, or a race to the top? Finally, accounting for the economic, environmental and other strategic interactions related factors that drive labeling incentives, what are the income distributional consequences of eco-labeling?

Existing studies on eco-labeling focus on the first question, and quantify the size of the green premium in various product markets either through consumer surveys, or hedonic price estimation.[1] More broadly, recent empirical studies on trade and the environment focus on how the relationship between trade liberalization and various environmental *outcome measures*, such as the intensities of air and water pollution, can be ascertained (Dean 2000, Antweiler, Copeland and Taylor 2001, Frankel and Rose 2002). Our approach here in this paper takes a different tact. Rather than focusing on the consumption end of the market for green products in which eco-labels are already in place, we begin instead by proposing the question, why do some countries have national eco-labeling programs and others do not? Meanwhile, our approach to uncover the link

---

[1] For instance, Robins and Roberts (1997) find that 5 to 15% of consumers may pay a slightly higher price for more environmentally friendly goods. A consumer survey in China indicated that close to 80% of consumers are willing to purchase green food (China Council for International Cooperation (1996). Also see Shams (1995) for the case of developing countries and Willer and Yuseffi (2001) for the case of eco-labeled apples in the United States. Also, Nimon and Beghin (1999) estimate the price premium for various individual attributes of apparel goods.

between eco-labeling and trade explicitly recognizes the endogeneity of environmental policy formation, and addresses the question of how the adoption of eco-labeling - a market-based environmental policy initiative - depends on a country's trade orientation, stage of development and other strategic concerns.

Following the analytical framework set out in Basu, Chau and Grote (2004), we consider a multi-country setting of export rivalry in two stages. In the first stage, countries determine whether or not to adopt eco-labeling. In the second stage, countries compete in a horizontally differentiated product market consisting of goods produced via a green production method and goods produced via a baseline production method. This frame-work yields a set of empirical implications in a subgame perfect Nash equilibrium, and highlights the selection criteria of countries that adopt eco-labeling. Consistent with recent empirical studies on the interlink between trade and the environment (Dean 2000, Antweiler, Copeland and Taylor 2001, Frankel and Rose 2002), it is shown that participation in world trade, the scale of production, and the stage of development of an economy are positively associated with the likelihood of eco-labeling. Taking these established results a step further, our findings also indicate the presence of strategic interdependence, in which the likelihood to adopt labeling is positively correlated with the popularity of labeling among a country's major export destinations. Thus, while the popular characterization of export competing countries as participants in a "race" may indeed be apt, the nature of such strategic interactions should perhaps be more appropriately termed a race to the top, rather than a race to the bottom.

In this paper, the findings of Basu, Chau and Grote (2004) are extended in two directions. In terms of analytics, a set of welfare consequences associated with the move towards eco-labeling by some countries and not others will be examined. By highlighting the endogeneity of labeling incentives directly, we find that the key lies not just in the size of the green premium, as is frequently alleged in the literature. Indeed, we will define an *industry-level* green premium in general equilibrium, which is key to the welfare consequences of export rivalry based on eco-labeling in a subgame perfect Nash equilibrium.

In terms of empirical analysis, an important question that has yet to be explored is whether the prior focus exclusively on export rivalry may have ruled out possible strategic interactions via import competition. Indeed, are labeling programs oriented towards foreign consumers' preferences in export markets, or are they possibly also instigated by the penetration of environmental friendly import competition?

In what follows, Section 2 presents a general equilibrium model that yields a set of possible determinants of the incentive to adopt eco-labeling.

Section 3 presents the welfare consequences of eco-labeling. The empirical methodology and the findings are presented in Section 4. Section 5 concludes.

2   The Basic Model

Producers in N countries are engaged in the production of two goods: a homogeneous numeraire Y^j, $j = 1,...,N$ and an agricultural output X^j. Production of the numeraire commodity employs a composite input, L^j_y, with $Y^j = \omega^j L^j_y$, where ω^j denotes the marginal and average product of input L^j_y.

Producers in agriculture also employ the composite input, and choose in addition between (i) an environmentally sound production technology X^j_e, or (ii) a baseline production technology X^j_o, with

$$X^j_e = (L^j_x / a)^\alpha, \qquad X^j_o = (L^j_x)^\alpha,$$

where $a \in (1, \infty)$ is producer-specific, and parameterizes the cost of adopting the environmentally friendly production technique. The cumulative distribution function X^j_e $F(a')$ denotes the fraction of producers in country j with $a \le a'$, and $a, a' \in (1, \infty)$. Let M^j be the number of competitive agricultural producers in country j.

2.1 Voluntary Adoption of Green Production Technique

Whether or not a producer in country j adopts the eco-friendly method of production is an outcome of a two-stage decision making problem, and depends in particular on the extent to which eco-labeling allows producers to internalize consumers' willingness to pay for eco-friendly production techniques. Let p_u be the price of unlabeled agricultural output produced via the baseline technique, and p^j_l be the price of labeled agricultural output produced via the environmentally friendly technology. We allow p^j_l to differ by country-of-origin, in order to account for the possibility that the green premium $(p^j_l - p_u)$ may differ across countries due to differing consumers' perception about the credibility of eco-labeling programs across countries, or simply due to differing consumers' perception about the location-specific environmental benefits, and hence, their willingness to pay for the implementation of green production techniques.

Each producer in agriculture first determines whether or not to voluntarily adopt the environmentally sound technology, and conditional on technology adoption choices, determines the amount of input L^j_x to employ. Beginning from the second stage, and taking as given the competitively determined cost of employing a unit of the composite input, ω^j, it is straightforward to verify that maximal profits respectively by choice of the environmentally sound technology, $\pi^j_e(a, p^j_l)$, and the baseline technology, $\pi^j_o(p_u)$, are given by:

$$\pi^j_e(a, p^j_l) = \max_{L^j_x} p^j_l(\frac{L^j_x}{a})^\alpha - \omega^j L^j_x \equiv (1-\alpha)\left(p^j_l(\frac{\alpha}{a\omega^j})^\alpha \right)^{\frac{1}{1-\alpha}}, \tag{1}$$

$$\pi^j_o(p_u) = \max_{L^j_x} p_u(L^j_x)^\alpha - \omega^j L^j_x \equiv (1-\alpha)\left(p_u(\frac{\alpha}{\omega^j})^\alpha \right)^{\frac{1}{1-\alpha}}, \tag{2}$$

Also, let $X^j_e(a, p^j_l)$ and X^j_o respectively denote the profit maximizing output levels associated respectively with equations (1) and (2). It follows, therefore, that a producer in country j benefits from adopting the environmentally friendly production method if and only if

$$\pi^j_e(a, p^j_l) \geq \pi^j_o(p_u) \Leftrightarrow a \leq \left(\frac{p^j_l}{p_u} \right)^{\frac{1}{\alpha}} \equiv \bar{a}^j.$$

In other words, the parameter $\bar{a}^j$ singles out the marginal producer who is just indifferent between the two techniques. Clearly, the higher the green premium, $(p^j_l/p_u)-1$, the higher will be the fraction of producers $F^j(\bar{a}^j)$ who benefit from green agricultural production.

The definition of $\bar{a}^j$ also implies that the value of aggregate agricultural production in country j is made up of two parts, derived respectively from environmentally friendly (X^j_e) and baseline (X^j_o) production:

$$M^j \left[p^j_l \int_{\underline{a}^j}^{\bar{a}^j} X^j_e(a, p^j_l)dF^j(a) + p_u \int_{\bar{a}^j}^{\infty} X^j_o(p_u)dF^j(a) \right] \tag{3}$$
$$= M^j [p_u(\frac{\alpha}{\omega^j})^\alpha]^{\frac{1}{1-\alpha}} \left[(1-F^j(\bar{a}^j)) + (\frac{p^j_l}{p_u})^{\frac{1}{1-\alpha}} \int_{\underline{a}^j}^{\bar{a}^j} (1/a)^{\frac{\alpha}{1-\alpha}} dF^j(a) \right].$$

As should be apparent, international differences in revenue per producer can be decomposed into two parts, including (i) terms in the first square brackets ($p_u(\alpha/(\omega^j)^\alpha)^{1/(1-\alpha)}$), which depend on international cost differences ω^j, and (ii) terms in the second square brackets, which depend on the self-selection among producers in employing the two agricultural production techniques ($\bar{a}^j$), and the green premium.

Note in particular that producer profits in countries where no eco-labeling programs prevail is in fact a special case of equation (3) above, in which p^{j}_{l} is replaced by p_u, as the green premium does not apply to unlabeled products. It follows, therefore, from the definition of $\bar{a}^{j}$ that $\bar{a}^{j} = (p^{j}_{l}/p_u)^{1/\alpha} = 1$. Thus, profits of the average producer simply depend on ω^{j} with:

$$\pi^{j} = (1-\alpha)(p_u(\frac{\alpha}{\omega^{j}})^{\alpha})^{\frac{1}{1-\alpha}}. \tag{4}$$

2.2   The Green Premium and Supply Response

Consumer preferences in country j are characterized by a utility function $(U^{j}(D^{j}_{x}, d^{j}_{y}))$, which accounts for consumption of the homogeneous numeraire d^{j}_{y}, and a consumption index of good x, D^{j}_{x}, with,[2]

$$\log U^{j}(D^{j}_{x}, d^{j}_{y}) = \beta^{j} \log D^{j}_{x} + (1 - \beta^{j}) \log d^{j}_{y},$$

where $\beta^{j} > 0$ denotes the share of consumer expenditure devoted to the consumption of the agricultural output. In addition,

$$D^{j}_{x} = \sum_{i=1}^{N}(1+g^{j})d^{ji}_{e} + \sum_{i=1}^{N}d^{j}_{o}$$

D^{j}_{x} is made up of two components, accounting respectively for the physical quantities of x consumed, $\sum_{i=1}^{N}d^{ji}_{e}$ and an index of green consumption $\sum_{i=1}^{N}g^{i}d^{ji}_{e}$. The ratio $(1+g^{i})/(1+g^{k})$ gives the marginal rate of substitution between d^{ij}_{e} and d^{ik}_{e} and reflects consumer's relative valuation for eco-friendly production originating from countries i and k. The marginal rate of substitution between a unit of labeled output from country i, and a unit of unlabeled output is simply $1+g^{i}$.

In equilibrium, relative prices must reflect these consumer preferences for there to be positive demand for all goods, and hence

$$\frac{p^{i}_{l}}{p^{k}_{l}} = \frac{1+g^{i}}{1+g^{k}}, \quad \frac{p^{i}_{l}}{p_u} = 1+g^{j}, i,k = 1,...,N. \tag{5}$$

---

[2] See Dixit and Stiglitz (1977) for a discussion of the use of similar utility indexes when product differentiation is of central concern.

It follows that aggregate agricultural producer revenue in the presence of eco-labeling in country j depends on the green premium, since:

$$Q_e^j(p_u) = M^j[p_u(\frac{\alpha}{\omega^j})^\alpha]^{\frac{1}{1-\alpha}}$$

$$\left[(1-F^j((\frac{p_l^j}{p_u})^{1/\alpha})) + (\frac{p_l^j}{p_u})^{\frac{1}{1-\alpha}} \int_{p_u}^{\frac{p_l^j}{p_u}^{1/\alpha}} (1/a)^{\frac{\alpha}{1-\alpha}} dF^j(a)\right]$$

$$= M^j[p_u(\frac{\alpha}{\omega^j})^\alpha]^{\frac{1}{1-\alpha}}$$

$$\left[1+(1+g^j)^{\frac{1}{1-\alpha}} \int^{(1+g^j)^{1/\alpha}} [(\frac{1}{a})^{\frac{\alpha}{1-\alpha}} - (\frac{1}{1+g^j})^{\frac{1}{1-\alpha}}]dF^j(a)\right]$$

$$\equiv p_u^{\frac{1}{1-\alpha}}\gamma^j(1+G^j).$$

In the absence of labeling, agricultural producer revenue in country j is given by:

$$Q_o^j(p_u) = p_u^{\frac{1}{1-\alpha}}\gamma^j.$$

with $\gamma^j \equiv M^j(\alpha/\omega^j)^{\frac{\alpha}{1-\alpha}}$. γ^j parameterizes the production cost in the agricultural sector of country j. Note also that $p_u G^j$ is an industry-level green premium, and represents the increase in industry-level revenue, holding p_u constant, that may be expected subsequent to eco-labeling. The size of G^j depends jointly on a demand-side and a supply-side effect. The demand effect is given by the country-specific green premium $1+g^j$, and G^j rises with g^j for every country j, with $G^j > 0$ if and only if $g^j > 0$. On the supply side, the cost distribution among producers in country j, F^j, matters, and a popular prevalence of producers at the lower end of the cost distribution implies a larger industry-level green premium.

2.3 Nash Equilibrium

In a Nash equilibrium, countries' decisions to adopt labeling are interdependent. We seek conditions under which a country will adopt eco-labeling in a Nash equilibrium, taking into account the endogenous terms of trade consequences of these adoption choices. To begin with, let I be the set of all countries in which an eco-labeling program is in place, and I_{-j} be the set of all countries in I but country j. With consumer income (aggregate earnings of composite input owners) equal to $\omega^j L^j$ in country j, aggregate world demand for the agricultural output is equal to total producer revenue if and only if:

$$\sum_{j=1}^{N}\beta^j\omega^j L^j =\sum_{j\in I}(p_u(I))^{\frac{1}{1-\alpha}}\gamma^j(1+G^j)+\sum_{j\notin I}(p(I))^{\frac{1}{1-\alpha}}\gamma^j.$$

It follows, therefore, that the price of unlabeled (eco-unfriendly) agricultural output is given by:

$$p_u(I)=\left(\frac{\sum_{j=1}^{N}\beta^j\omega^j L^j}{\sum_{j\in I}\gamma^j G^j +\sum_{j=1}^{N}\gamma^j}\right)^{\frac{1}{1-\alpha}}.$$

Note that the price of eco-unfriendly products is strictly decreasing in the number of countries that have instituted an eco-labeling program, as long as $G^j > 0$ for $j \in I$. Indeed, the same is true of the price of labeled products, since $p^j_l(I, g^j) = p_u(I)(1+g^j)$. These terms of trade effects accordingly highlight the negative externality that one country's decision to implement labeling programs imposes on the welfare of producers in other countries.

What remains to be seen, however, is how the decision to adopt eco-labeling in one country depends on that of another. To this end, let W be the sum total of consumer expenditure in the N countries to be devoted to the consumption of the agricultural output, with $W \equiv \sum_{j=1}^{N}\beta^j\omega^j L^j$. Aggregate producer profits in country j with eco-labeling, taking as given the I_{-j}, is given by:

$$\Pi_e^j(I_{-j},G^j)=\frac{(1-\alpha)W\gamma^j(1+G^j)}{\sum_{i\in I_{-j}}\gamma^i G^i +\gamma^j G^j +\sum_{j=1}^{N}\gamma^j}. \tag{6}$$

In contrast, if country j abstains from encouraging green production techniques via eco-labeling, aggregate producer profits in country j is equal to:

$$\Pi_o^j(I_{-j}) = \frac{(1-\alpha)W\gamma^j}{\sum_{i\in I_{-j}}\gamma^i G^i + \sum_{j=1}^{N}\gamma^j}. \tag{7}$$

Thus, if c^j denotes the fixed cost required to put in place a credible labeling program in country j, aggregate producer profits rise with market-based voluntary green production via eco-labeling, taking as given the adoption decisions of the rest of the N -1 countries, if and only if $\Pi_e^j(I_{-j}, G^j) - \Pi_o^j(I_{-j}) \geq c^j$, or[3]

$$\log G^j \geq \log(\frac{c^j/(1-\alpha)}{Q_o^j(p(I_{-j}))}) -$$
$$\log(1 - \frac{c^j/(1-\alpha)}{W - Q_o^j(p(I_{-j}))}) + \log(1 + \frac{\gamma^j}{\sum_{i\in I_{-j}}\gamma^i G^i + \sum_{i\neq j}\gamma^j}). \tag{8}$$

As such, the decision to implement an eco-labeling program reflects a number of factors that are simultaneously in play. To begin with, the larger the industry-level green premium G^j, the more likely it is that the inequality in equation (8) is satisfied. In addition, the value of aggregate output of country j, Q_o^j (p_u (I_{-j})), also plays a key role in the determination of labeling incentives. First, the larger the output level in the absence of an eco-labeling program in country j, Q_o^j (p_u (I_j)), the more able are producers in country j in shouldering the fixed cost of labeling. However, and contrary to the first effect, a country that has a sufficiently large market share to begin with may also have little to gain from market share rivalry via eco-labeling. To see this, note that if country j is large enough so that W - Q_o^j (p_u) is close to zero, $\Pi_e^j(I_{-j}, G^j)$ - $\Pi_o^j(I_{-j})$ - c^j is always less than zero, for $c^j > 0$.

The third term in equation (8) denotes the magnitude and the nature of peer effects between the N countries. In particular, linearizing

---

[3] To see this, note from equations (6) and (7), along with the definition Q_o^j (p_u), that π_e^j (I_{-j}) - π_o^j (I_{-j}) > c^j if and only if c^j / Q_o^j (p_u)
$\leq G^j(\sum_{i\in I_{-j}}\gamma^i G^i + \sum_{i\neq j}\gamma^j)/(\sum_{i\in I_{-j}}\gamma^i G^i + \gamma^j G^j + \sum_{i=1}^{N}\gamma^i)$.

Equation (8) follows from rearranging terms, and taking logs on both sides.

$$\log(1+\frac{\gamma^j}{\sum_{i\in I_{-j}}\gamma^i G^i +\sum_{i\neq j}\gamma^j}) \text{ with respect to } \gamma^j/\sum_{i\neq j}\gamma^j, \text{ we obtain}$$

$$\log(1+\frac{\gamma^j}{\sum_{i\in I_{-j}}\gamma^i G^i +\sum_{i\neq j}\gamma^j}) \approx \log(1+\frac{\gamma^j}{\sum_{i\neq j}\gamma^j}) - \frac{\gamma^j}{\sum_{i\neq j}\gamma^j}\left(\sum_{i\in I_{-j}}\frac{\gamma^i}{\sum_{i=1}^N \gamma^i}G^i\right).$$

Among other things, adoption is more likely: (i) as the cumulative number of countries that have already adopted a labeling program I_{-j} increases, so long as $G^i > 0$, and (ii) as the industry-level green premium of those countries $\gamma^i G^i$ that already have a labeling program in place also increase. In addition, the comparative production cost advantage of country j in baseline agricultural production ($\gamma^j/\sum_{i\neq j}\gamma^j$) can have a positive impact on labeling incentives, so long as the industry green premium of country j's export rivals ($\sum_{i\in I_{-j}}\gamma^i G^i$) is sufficiently large.

Notably, the cumulative number of countries with labeling programs plays a role in adoption decisions *only if* the industry level green premia of the exports of these same countries are strictly positive. In addition, a presumption in popular discussions on the potential detrimental effect of eco-labeling on market access is that developing countries bear a dispro-portionate disadvantage with eco-labeling precisely because the industry level green premium is smaller for developing countries. This may be due to the possibilities that: (i) consumers attach a smaller premium to labeled products from developing countries (a smaller g^i); and / or (ii) producers in developing countries have an inherent disadvantage in producing the environmentally friendly output (F^i of a developing country stochastically dominates F^i of a developed country). From the definition of G^i, both of these possibilities can contribute to a reduction in the industry level green premium. In the context of our analysis, therefore, equation (8) also opens up a way of testing whether these allegations apply, by examining whether developed and developing countries exert differential influence on the adoption behavior of countries that have yet to adopt eco-labeling. Proposition 1 summarizes these observations:

Proposition 1:

In a Nash equilibrium, the incentives to adopt a voluntary eco-labeling program in country j depends systematically on:

1. the fixed cost of eco-labeling;

2. a scale effect that is represented by the size of existing output prior to labeling;

3. the comparative production cost advantage of country j in the baseline technique of production $\gamma^j / \sum_{i \neq j} \gamma^i$,

4. peer effects as determined by the number of other countries that have already implemented an eco-labeling program, and the industry-level green premium of these countries.

3 Welfare Implications

We now turn to the welfare implications of eco-labeling. In any Nash equilibrium with export rivalry based on eco-labeling, two sets of countries can be identified. The first group includes a Nash equilibrium set $\widetilde{I}$ of countries that willingly incur the fixed cost c^j and implement an eco-labeling program, with

$$\Pi_e^j(\widetilde{I}_{-j}, G^j) - \Pi_o^j(\widetilde{I}_{-j}) \geq c^j$$

Meanwhile, a second group of countries are characterized by the lack of incentives to adopt labeling, since

$$\Pi_e^j(\widetilde{I}_{-j}, G^j) - \Pi_o^j(\widetilde{I}_{-j}) < c^j$$

In what follows, the welfare comparison conducted takes the case where no country adopts eco-labeling as a baseline. We evaluate the welfare of the two groups of producers enumerated above, along with the welfare of the representative consumer in a Nash equilibrium wherein at least one country adopts eco-labeling.

3.1 Aggregate Producer Welfare Implications

From the definition in equation (6), for all country $i \notin \widetilde{I}$, aggregate producer profits are given by

$$(1-\alpha)(p_u(\widetilde{I}))^{\frac{1}{1-\alpha}} \gamma^i$$

Thus, aggregate producer profits necessarily decline, relative to a regime in which no country adopts eco-labels, via a terms of trade effect that impacts on the price of unlabeled products. In particular, the higher the Nash equilibrium number of countries that have adopted eco-labeling, the larger will be the profit reduction facing producers in this group.

For countries that do adopt eco-labeling in a Nash equilibrium, however, the aggregate producer profits derived from eco-labeling depend jointly on the terms of trade effect, and the country-specific industry level green premium. To see this, recall that aggregate producer profits are given by

$$(p_u(\widetilde{I}))^{\frac{1}{1-\alpha}} \gamma^i (1+G^i)$$

Making use of the equilibrium price level $p_u(\widetilde{I})$, it is straightforward to verify that country j is strictly better off *only if*

$$G^j > \sum_{i \in I_{-j}} \frac{\gamma^i}{\sum_{i \neq j}^N \gamma^i} G^i.$$

Thus, even if incentives are right for a country to engage in labeling, aggregate producer profits may still decline relative to a regime in which no country adopts eco-labeling. In particular, aggregate profits increase only for a subset of countries with a sufficiently high industry-level green premium.

3.2   Individual Producer Welfare Implications

While the discussion above focuses on the country-level producer welfare implications of eco-labeling in a Nash equilibrium, a similar comparison can be conducted by focusing on the impact of eco-labeling on individual producers. In particular, since individual producer profits in the absence of labeling are given by:

$$\pi_o^j(p_u(\widetilde{I})) = (1-\alpha)\left(p_u(\widetilde{I})(\frac{\alpha}{\omega^j})^\alpha \right)^{\frac{1}{1-\alpha}},$$

it follows that producers in any country j who do not adopt environmentally sound production techniques (with $a \geq \overline{a}^j$), and therefore cannot take advantage of the green premium made available via eco-labeling, are necessarily worse off. These profit losses are a direct consequence of the price decline subsequent to the adoption of eco-labeling by any country.

Meanwhile, for the rest of the producers who voluntarily adopt environmentally sound production technologies, their profits in a Nash equilibrium are given by

$$\pi_e^j(a, p_l^j) = (1-\alpha)\left[p_l^j (\frac{\alpha}{a\omega^j})^\alpha \right]^{\frac{1}{1-\alpha}}.$$

It follows that the impact of eco-labeling on the profits of these producers depends once again on the joint impact of a terms of trade effect through a reduction in $p_u(\widetilde{I})$, along with the green premium g^j. In particular, producers in country j who adopt the environmentally friendly production technique are strictly better off if and only if

$$g^j > \sum_{i \in I_{-j}} \frac{\gamma^i}{\sum_{i \neq j}^N \gamma^i} G^i.$$

3.3 Aggregate Consumer Welfare Implications

Finally, turning to the impact of eco-labeling on the welfare of consumers, we note that the indirect utility of a consumer in country j (with labor income ω^j) can be expressed as

$$\log \omega^j - \beta^j \log(p_u(\widetilde{I})) + K,$$

where $K \equiv \beta^j \log \beta^j + (1-\beta^j) \log(1-\beta^j)$ is a constant. It follows, therefore, that since eco-labeling decreases the price of unlabeled products, consumer welfare strictly improves in each country i as long as at least one country adopts a labeling program in a Nash equilibrium. There are a number of other possible considerations that may be incorporated into the basic analysis, including import taxes, or the share of fiscal burden of the labeling program. These are discussed in detail in Basu, Chau and Grote (2003). However, the main thrust of this finding remains robust.

4   Empirical Analysis

National eco-labeling programs for agricultural products can be found in most OECD countries but also increasingly in many developing countries (Conway 1996). In the agricultural and food industry sector, certification refers to a wide array of food products (juices, cereals and grain including rice, and even alcoholic beverages, sugar, meat, dairy products or eggs) produced either by organic or bio-dynamic farming technologies or through integrated pest management (FAO 2000). Certification can also refer to agricultural food and non-food products (coffee, tea, cocoa, and flower) which are produced with less fertilizers and pesticides as opposed to traditional practices on plantations and in monoculture. Also, other non-food agricultural products like animal feeds (for production of organic meat, dairy products and eggs), grain seeds, natural pesticides and insecticides, cosmetics and textiles (cotton, leather and leather goods) may also be certified if they meet certain environmental criteria.

In this section, we present an empirical approach to answer the three questions enumerated at the outset of this paper. Specifically, we are interested in determining whether there exists a competition-induced limit to the threat of environmental exploitation in the face of increasing international trade. In particular, does the export orientation of a country determine at least in part its decision to adopt environmentally friendly production technologies via eco-labeling? In addition, we will approach the question of whether there is a race to the bottom by examining how the cumulative adoption of eco-labeling by other countries affects the incentives to adopt by developed and developing countries alike. Finally, by uncovering the potential determinants of eco-labeling adoption, we can infer the potential welfare impacts of eco-labeling on developed and developing countries, based on our findings elaborated in section 3.4.

A key issue is thus how observed incidences of eco-labeling may reveal information on producers' perception of the size of the industry level green premium. To this end, we refer to the right hand side of equation (8), which suggests the inclusion of the following regressors to capture and control for (i) the cost of eco-labeling; (ii) scale effects; (iii) production cost and (iv) peer effects.

The eco-labeling data on which our empirical analysis is based is described at length in Basu, Chau and Grote (2004). It tracks the prevalence of national level food industry eco-labeling initiatives from 1976 to 1999 in 66 countries, and if present, the time of adoption. Of the 66 countries, 30 countries have instituted an eco-labeling program by the end of 1999, about two-thirds are developing economies, and about half are on average net food industry importers (exporters) over the course of 1976 - 1999. In addition, we assembled macroeconomic, bilateral trade, and food industry environmental performance data for these countries. Summary statistics are reproduced in Table 1. To capture the fixed (administrative) cost of eco-labeling, c^j is taken to depend on: (i) the stage of development of an economy -- real gross domestic product per capita, (World Bank 2001b) and (ii) the existing level of food industry environmental damage -- average pre-labeling food industry water pollution (share of total BOD emission) (World Bank 2001b)[4]. To capture scale effects, we have assembled data on the average pre-labeling food industry total output share of the 66 countries, and these are taken from World Bank (2001a). The comparative cost advantage of country j is proxied by the export orientation of the economy -- the average pre-labeling share of total exports to total food industry trade (Trade and Production Database, World Bank 2001a).

While Table 1 presents unconditional comparisons, and does not properly control for the joint impact of all of these variables on adoption decisions, it paints a picture that is largely consistent with that of equation (8). In terms of fixed cost, Table 1 shows that developed countries, and countries with relatively low levels of food-industry pollution, appear to be more capable of bearing the cost of instituting a labeling program. In terms of scale effects and cost differences, Table 1 also shows that countries with higher output levels and a comparative cost advantage appear to be favorably selected in the set of countries with eco-labeling programs.

With respect to peer effects, and to uncover the impact of trade competition on eco-labeling, we consider two types of peer groupings in this paper: "wcexdest^j_t" and "wcimori ^j_t". As in Basu, Chau and Grote (2004), the variable "wcexdest ^j_t" is constructed for country j at time t by computing the weighted cumulative number of countries in the peer group other than country j that have adopted eco-labeling since 1976 till time t-1. The weights are taken to be the food industry output $\gamma^j / \sum_{i=1}^{N} \gamma^i$ (equation (8)) of country j as a share of the total output of the 66 countries. For example, to compute the peer effect based on bilateral export competition at any time t, food industry bilateral trade data is employed to identify top ten export destinations for each country j. The weighted cumulative number of these export destinations that have an eco-labeling program in place up till time t-1 for each country j gives "wcexdest^j_t" at time t. In order to consider the possible impact of import competition on eco-labeling incentives, we construct an analogous variable "wcimorit^j_t". Here, the relevant peer group is the top ten import origins of food industry imports for each country j. The weighted cumulative number of the country's top ten import origins that have an eco-labeling program in place up till time t-1 for each country j gives "wcimorit^j_t".

---

[4] All averages used as regressors in our estimation are computed for years that fall between 1976 - 1999, but prior to (and not including) the year during which eco-labeling is adopted for each country.

Table 1. Trade links and output pre and post eco-labeling

Variable	Pre*	sd	Post**	sd	NA***	sd
Real per capita Income (US$ 1995 const.)	10299.750	9400.600	13505.990	12103.420	3487.938	4639.820
Food ind. export orientation (% export to total food ind. exports & imports)	60.157	21.953	56.469	22.628	48.426	25.143
Food ind. export share to US, WE & JPN (% export to US, WE, JPN to total export)	53.378	21.001	56.622	20.593	47.567	24.983
Food ind. output share (% total world output)	3.234	6.927	3.602	6.997	0.222	0.250
Food ind. water-pollution (% total BOD emission)	44.203	10.298	43.707	9.955	51.046	16.056

*Mean country annual averages during the pre-labeling periods for countries that instituted a labeling program after 1976.
**Mean country annual averages during the post-labeling periods for countries that instituted a labeling program after 1976.
***Mean country annual averages from 1976 to 1999 for countries that never instituted a labeling program.
Source: Basu, Chau and Grote (2004).

Finally, in order to examine the possibility that the industry level green premium of developing countries, and hence their impact on the labeling incentives of other trading partners, may be significantly different from that of their developed country counterpart, we construct two additional peer effect variables: "wcdevexdestt^j_t" and "wcdevimorit^j_t". Respectively, these are the weighted cumulative number of developing country trade partner (export destinations and import origins) of country j that instituted a labeling program from 1976 up until time t-1.

4.1 Estimation Results

We take the approach of estimating a proportional hazard model. Let x_{jt} be a vector of time-varying explanatory variables, where $t = 1976,...,T_j$, when country j implements an eco-labeling program.

The hazard rate at t_j -- the probability of adoption when t_j years have passed given that adoption has not yet taken place -- is simply

$$h(t_j \mid \mathbf{x}_{jt}) = \frac{v'(t_j \mid \mathbf{x}_{jt})}{1 - (v(t_j \mid \mathbf{x}_{jt}))}.$$

We assume a model with proportional hazard (Cox 1972), and specify in addition that each of the K time-varying covariates enter into the determinant of the hazard rate as follows:

$$h(t_j \mid \mathbf{x}_{jt}) = \hat{h}(t_j) e^{\sum_{k=1}^{K} \beta_k x_{jkt}}.$$

where $\hat{h}$ denotes the baseline hazard function. Thus, $e^{\beta_k} \geq (<)1$ represents the hazard ratio for a unit change in x_{jkt}. These estimates are obtained by maximizing a partial log-likelihood function (Kalfleisch and Prentice 1980). Since only data prior to adoption will be used, the problem of endogeneity of x_{jkt} subsequent to labeling does not arise. In addition, the estimation procedure does not place restriction on the unknown functional form of the baseline hazard function.

Table 2. Proportional hazard regression

Hazard Ratios	I	II	III	IV
Real per capita				
Income	1.00007 ***	1.00007 ***	1.00008 ***	1.00008 ***
	0.00002	0.00002	0.00002	0.00003
food ind. output share	1.36124 ***	1.38671 ***	1.36086 **	1.38472 ***
	0.16898	0.17345	0.17665	0.18058
food ind. output share				
-squared	0.99256 **	0.99228 **	0.99213 **	0.99185 **
	0.00362	0.00362	0.00381	0.00381
food ind. water				
pollution	0.95269 ***	0.95018 ***	0.94623 ***	0.94340 ***
	0.01728	0.01785	0.01711	0.01782
food ind. exports				
share	1.05236 ***	1.05232 ***	1.05604 ***	1.05536 ***
	0.01615	0.01917	0.01676	0.02012
wcexdest		1.01866 *		1.02306 ***
		0.01087		0.01135
wcdevexdest		0.91896 **		0.90885 ***
		0.03870		0.03917
wcimori			0.96072	0.95070 *
			0.02412	0.02500
wcdevimori			0.63903	0.66809
			0.22910	0.23382
No. of observations	1089	1089	1089	1089
Incidences of eco-				
labeling	21	21	21	21
Log Likelihood	-63.249	-62.488	-61.423	-60.460
Wald chi^2	35.220	34.440	32.730	39.900
Prob > chi^2	0.000	0.000	0.000	0.000

Robust standard errors (Lin and Wei, 1989) in parenthesis.

*significant at the 10% level

**significant at the 5% level

***significant at the 1% level.

Table 2 presents our findings. The first column replicates the result of Basu, Chau and Grote (2004),[5] and shows in particular that a higher real per capita GDP, and a lower existing level of food industry related water pollution are both associated with high likelihood of eco-labeling, as the estimated coefficients β are strictly greater than one, and significant at the 1% or 5% level.

The scale effect figures prominently as well, having a significant and positive impact on the likelihood of eco-labeling. The results also indicate that scale effects are nonlinear, in that the rate of increase in the likelihood of eco-labeling decelerates with scale. In addition, the export orientation of the food industry matters, and the likelihood of eco-labeling is positively associated with the share of food industry export to total food industry trade.

Finally, with respect to the two types of peer effects, the bilateral export destination peer effect continues to be strictly greater than unity and significant, whereas the import competition peer effect is not significant. These may be interpreted as an indication of *strategic complementarity* between countries, and particularly those that are engaged in export competition. Our findings also lend support to the importance of labeling as an export promotion device, rather than an import deterring instrument. The estimated coefficient on developing country peer effects "wcdevexdest" is significant and less than one, with the interpretation that for controlling for other factors, the influence that a developing country may have on Nash equilibrium labeling initiatives is indeed smaller. In the context of our theoretical discussion, one possible reason behind this could be that the industry green premium of developing country exports is small relative to their developed country counterparts.

We believe that we have merely taken the first steps to examine whether developing countries may indeed be subject to reputational / technological disadvantage in green production compared to developing countries. Further research based, for example, on eco-labeling initiatives in other

---

[5] For each of these estimations, we report the number of observations, the number of incidences of eco-labeling that took place after 1976, the log likelihood and Wald Chi-squared statistics of the estimation. The hypothesis that all of the estimated coefficients are all equal to one is rejected in all of our estimations, at a significance level of less than 1%. Also, note that the number of incidences of eco-labeling applicable in the estimation is 21, as comparable pre-labeling data on output share, trade orientation and the like are not available to us. As shown, nine countries (Austria, Finland, France, Germany, Italy, New Zealand, Norway, Sweden and the United Kingdom) instituted their eco-labeling programs prior to 1976. This leaves a total of 57 countries that are included in our estimations.

product markets, or using other plausible peer effect variables as in Basu, Chau and Grote (2004), may well shed light on the pervasiveness of reputational concerns and technological disadvantage unveiled in our findings here.

5   Conclusion

In the context of the role of voluntary and market-based policy instruments that elicit environmentally friendly production practices, as well as popular concerns regarding the threat of environmental exploitation in the face of increasing international trade, this paper raised a number of questions. First, we ask whether market incentives made available through eco-labeling entice countries engaged in export competition to improve environmental performance? Second, how do countries engaged in trade competition interact with one another when the strategic variable in question is the need to establish reputational comparative advantage in a segmented market where consumers have a choice between products manufactured via environmentally friendly and environmentally unfriendly means (Basu and Chau 1998)? Finally, is there a development and environment trade-off when countries compete based not just on comparative cost advantages, but also on their ability to shoulder the costs associated with a credible eco-labeling program?

Based on the model developed in Basu, Chau and Grote (2004), we set out to comparing producer welfare with and without competition based on eco-labeling. The key, as it turns out, lies in the size of the industry level green premium, which depends on both (i) a demand side consumption green premium effect, and (ii) a supply side cost of green technological adoption effect. Interestingly, we find that in the absence of international coordination of technology adoption, and an appropriate way in which the gains from eco-labeling can be shared, countries that find themselves voluntarily engaged in eco-labeling initiatives in a subgame perfect Nash equilibrium are not necessarily made better off. Meanwhile, countries that opt out are worse off because of the terms of trade effect of eco-labeling on products made using the baseline technology.

In terms of empirical findings, this paper takes the result of Basu, Chau and Grote (2004) a step further, and examines the extent to which eco-labeling should be viewed as an export promotion device, or an import deterring mechanism. Our findings based on our construction of the import peer effect variable is in favor of the former, with the peer effect variables with peer grouping based on the degree of export competition showing up again as significant and positive, though the peer effect based on import penetration is not significant.

Taken together, these findings indicate a set of possible answers to the three questions posed at the outset of this paper, and suggest additional research questions. To begin with, while production specialization induced by international trade may encourage environmental exploitation when the exportable industry is pollution intensive, our findings suggest that the export orientation of an industry can itself be a driving force that makes the practice of eco-labeling an attractive option. Meanwhile, with strategic complementarity in adoption as shown in the theoretical and empirical discussions of this paper - at least insofar as eco-labeling in the food industry is concerned - our findings suggest that a race to the top may in fact be in play.

While this paper has focused on the determinants of eco-labeling, a natural course for future research will clearly be to determine the consequences of eco-labeling, in terms of the greening of agriculture, welfare and market access. What the findings of this paper suggest in terms of research strategy, however, is that eco-labeling is far from an exogenous event. Rather, the adoption of eco-labeling is itself conditional on environmental performance, the stage of development of a country, and trade-related factors.

References

Antweiler W, Copeland BR, Taylor MS (2001) Is free trade good for the environment? American Economic Review 91:877-908

Basu AK, Chau NH (1998) Asymmetric country-of-origin effects on intra-industry trade and the international quality race. Review of Development Economics 2:140-166

Basu AK, Chau NH (2001) Market access rivalry and eco-labeling standards: are eco-labels non-tariff barriers in disguise? Department of Applied Economics and Management Working Paper, Cornell University, WP 2001-15

Basu AK, Chau NH, Grote U (2003) Eco-labeling and stages of development. Review of Development Economics 7:228-247

Basu AK, Chau NH, Grote U (2004) On export rivalry, eco-labeling and the greening of agriculture. Agricultural Economics 31:135-147

Conway T (1996) ISO 14000 standards and China: a trade and sustainable development perspective. International Institute for Sustainable Development (IISD); Prepared for the China Council for International Cooperation on Environment and Development, Working Group on Trade and Development (1996) China's green food development and environmental protection. http://www.iisd.org/pdf/isochina.pdf

Copeland BR, Taylor MS (1994) North-south trade and the environment. Quarterly Journal of Economics 109.755-787

Copeland BR, Taylor MS (1995) Trade and transboundary pollution. American Economic Review 85:716-737

Cox DR (1972) Regression models with life-tables. Journal of the Royal Statistical Society Series 34:187-220

Dean JM (2000) Does trade liberalization harm the environment? A new test. CIES Policy Discussion Paper #0015, University of Adelaide, Centre for International Economic Studies, Adelaide

Dixit A, Stiglitz JE (1977) Monopolistic competition and optimum product diversity. American Economic Review 67:297-308

FAO (2000) Food safety and quality as affected by organic farming. Paper (ERC/00/7) for the Twenty Second FAO Regional Conference for Europe in Porto, Portugal, 24-28 July 2000

Frankel JA, Rose A (2002) Is trade good or bad for the environment? Sorting out the causality. NBER Working Papers no 9201 National Bureau of Economic Research

Grossman GM, Krueger AB (1995) Economic growth and the environment. Quarterly Journal of Economics 110:353-77

Kalfleisch JD, Prentice RL (1980) The statistical analysis of failure time data. John Wiley & Sons, New York

Lin DY, Wie LJ (1989) The robust inference for the cox proportional hazards. Model Journal of the American Statistical Association 84:1074-1078

Nimon W, Beghin J (1999) Are eco-labels valuable? Evidence from the apparel industry. American Journal of Agricultural Economics 81:801-811

Porter ME, van der Linde C (1995) Toward a new conception of the environment-competitiveness relationship. Journal of Economic Perspectives 9:97-118

Robins N, Sarah R (ed) (1997) Unlocking trade opportunities case studies of export success from developing countries. International Institute for Environment and Development for the UN Department of Policy Coordination and Sustainable Development, London

Shams R (1995) Eco-labelling and environmental policy efforts in developing countries. Intereconomics 30:143-149

UNDP (1999) Human development database 11. UNDP, New York

UN (1997) Trade effects of eco-labelling. Proceedings of a seminar held in Bangkok 17-18 February, United Nations, New York

Willer H, Yussefi M (2001) Ökologische agrarkultur weltweit 2001. Stiftung Ökologie und Landbau, Bad Dürkheim

World Bank (2001) Trade and production database. The World Bank, Washington DC. http://www.worldbank.org/research/trade

World Bank (2001) World development indicators 2001. The World Bank, Washington DC

Science, Opportunity, Traceability, Persistence, and Political Will: Necessary Elements of Opening the U.S. Market to Avocados from Mexico

David Orden and Everett Peterson

1   Introduction

Technical barriers are often significant obstacles to market access for agricultural exporters. One approach to easing such technical trade restrictions is to shift from most restrictive instruments such as complete bans to less restrictive instruments of pest control. The key to such an alternative is often a systems approach to risk management, whereby a set of procedures are specified that reduce the pest-risk externality associated with trade of a commodity. The system measures add to exporter production costs but enable market access to occur. Adoption of systems approaches rest on a firm foundation in Article 5.6 of the WTO SPS Agreement which states that Members shall ensure that their measures "are not more trade-restrictive than required to achieve their appropriate level of sanitary or phytosanitary protection" (WTO 1994; Josling et al. 2005).

Since 1997, a long and contentious dispute between Mexico and the United States over U.S. restrictions on importation of Hass avocados has been partially resolved by replacing an import ban with trade under a system of risk mitigation measures. This case illustrates that progress can be made in easing technical trade restrictions - at least when the risk issues can be sharply delineated and addressed and governments are firmly committed to the negotiations. Easing of the longstanding import ban on Mexican avocados is trade-facilitating progress that has opened the U.S. market to Mexican producers in successive steps.

2   Background

2.1   The Avocado Quarantine

The ban on imports of Mexican avocados was promulgated in 1914 when there were no known controls (chemical or natural predators) for certain host-specific avocado pests prevalent in Mexico but not present in the United States. Subsequent development of modern pesticides and cultural practices has allowed the Mexican state of Michoacan to establish an industry of approved export-oriented avocado orchards. These orchards have successfully met the pest control standards of countries such as Canada and Japan, where avocados are not grown but there are potential concerns about transmission of fruit fly infestations. Mexican quarantine authorities have argued that the Michoacan avocado export protocols also provide adequate protection against pest risks of U.S. concern: that the region has low incidence of pests of quarantine significance, that the Hass avocado is not a host, or at least not a preferred host, for fruit flies, and that a systems approach to handling fruit for export has proven effective in eliminating risks of pest infestations being carried abroad. Mexico has contended that the U.S. ban cannot be justified on a risk basis, but was maintained to protect the U.S. industry economically. The U.S. avocado industry, concentrated in southern California, bitterly opposed opening the U.S. domestic market to Mexican avocados. The industry acknowledged that it receives prices well above those of Mexican exports, but asserted that it fears pest infestations associated with trade not competition in the marketplace. Domestic U.S. producers challenged Mexican assessments of pest risks and the effectiveness of the systems approach to risk management.

Caught in the middle of this controversy has been the U.S. Department of Agriculture. Twice during the 1970s USDA took preliminary steps to ease the avocado import ban, but in both cases the decision was aborted.[1] The issue lay unresolved through the 1980s, until NAFTA negotiations started in 1991 provided an opportunity for Mexico to raise its concerns again. Avocados dominated the agenda of many meetings of a joint Phytosanitary Working Group, where scientists from USDA's Animal and Plant Health Inspection Service (APHIS) and Mexico's Direccion General de Sanidad Vegetal (DIGSV) sparred over data requirements, research design, and interpretation of research results concerning possible lifting of the import ban. The technical debates centered on assessment of pest popula-

---

[1] Roberts and Orden (1996) provide a detailed analytic chronology of the avocado dispute.

tions, the host status of Hass avocados for fruit flies, and the adequacy of various proposed pest-risk mitigation strategies.

It took four years of bi-lateral procedural negotiations, data collection and analysis before USDA agreed to consider a Mexican plan for easing the avocado quarantine under a systems approach to pest risk mitigation. With some further safeguards, a proposed rule was published by USDA in July 1995 to allow imports of Mexican avocados grown and processed under specified conditions (USDA 1995). The proposed systems approach required annual surveys to determine pest incidence and pre-harvest, harvest, transport, packing, and shipping measures designed to reduce pest risks.[2] The distribution of imports was to be further limited to the northeastern United States, to avoid geographic proximity with regions susceptible to pest risks, and to four winter months when the risk of establishment of pests was mitigated by adverse weather.[3] Traceability was required, with identification required so that any infested fruit detected through inspections could be tracked back to the orchard from which it originated. USDA concluded that its proposed approach would provide an adequate level of security to domestic growers. Overall, USDA reported that with the proposed systems approach in place a seed pest or fruit fly outbreak was estimated to occur on average less than once every 1,000,000 years and a stem weevil outbreak might occur on average once every 11,402 years.

2.2   Domestic Opposition to Change

With the geographic and seasonal restrictions in USDA's proposed rule, partial easing of the ban opened less than 5% of the annual U.S. market to Mexican avocados. Even this partial access was fought aggressively by the domestic industry. The opposition was coordinated by the California Avocado Commission (CAC), which had closely monitored the deliberations from the outset of the NAFTA negotiations. The industry made the argument that the avocado quarantine should not be sacrificed to the political imperative of achieving a trade agreement. This was an aggressive strategy by the industry that turned on its head the conventional perception that regulatory processes are often under excessive pressure not from

---

[2] Pest of concern were identified as avocado-specific (three seed weevils, one stem weevil and one seed moth) and non-specific (three fruit flies).

[3] The region referred to as the northeastern United States or northeast in this paper includes two regions often separated in avocado shipment data: the northeast and east central regions. Mexican avocados were allowed into Alaska starting in 1993.

foreign but from domestic interest groups. Numerous declarations were made by the U.S. growers to the effect that "science might be traded off in a rush to sign a trade deal."[4]

The CAC argument was that imports of Mexican avocados under the proposed systems approach posed an unacceptable risk of pest infestation to domestic groves. The industry asserted that the surveys of pest incidence had failed to establish low population levels in the Michoacan growing area, that the proposed monitoring protocols were inadequate, and that Hass avocados were a better host of fruit flies than Mexico acknowledged. Technical criticism of the pest surveys were detailed, including, for example, objections to incorrect trap placement, weak trapping bait, insufficient climatological records, and inadequate trapping densities.[5] Any infestations of domestic groves that resulted from impor-tation of Mexican avocados would be costly to contain due to U.S. pesticide regulations and the close proximity of the domestic groves to residential neighborhoods. Thus, the CAC recommended that Mexico should be allowed to export avocados only under stringent conditions: that it could establish pest-free zones, that the imported avocados were treated with a pesticide which assured at a very high probability level that exotic pests were eliminated, or that additional scientific research unequivocally established that Hass avocados were not hosts of pests which are injurious to avocados and other fruits and vegetables grown in the United States.[6]

The conditions specified by the CAC for amendment of the avocado quarantine could effectively have precluded importation of Hass avocados from Mexico. The first condition, establishing and maintaining a pest free zone, required substantial eradication, monitoring, and quarantine enforcement costs well beyond the perimeters of commercial export groves in Mexico. Although it might eventually prove feasible technically, such an approach was regarded as uneconomical by Mexican officials who believed pest risks were already negligible. On the second condition, all parties agreed that no adequate post-harvest treatment was available. The third condition, strictly interpreted, also could not be met. The results of DIGSV's fruit fly host status research had indicated that fruit flies will attack Hass avocados shortly after they have been harvested. It was anticipated that additional research to rigorously establish the host status of Hass avocados would confirm that they are non-preferred hosts, but not the higher standard of "unequivocal non-host" that the CAC recommended.

---

[4] "Free Trade with Mexico" Betsey Blanchard Chess, California Grower, 6/91, p19
[5] Statement by the California Avocado Commission, Docket No. 94-116-1, 1/3/95.
[6] Statement by the CAC for Docket No. 94-116-1, ANPR Concerning the Importation of Fresh Hass Avocado Fruit Grown in Michoacan, Mexico, 2/95, p 2

Industry opposition orchestrated by the CAC was effective in temporarily blocking change to the quarantine when USDA announced it would not make a decision on a final rule to allow avocado imports in time for the 1995-96 winter shipping season. The CAC kept up its pressure in 1996. It threatened legal action to block lifting of the ban and attempted to circumscribe USDA authority through an amendment to congressional appropriations legislation for APHIS. Full-page advertisements were placed in several national newspapers by the CAC. Against the backdrop of a hangman's noose or smoking gun, these ads claimed that "The USDA is about to sign the death warrant for a billion dollar American industry."[7] The CAC also filed a new petition with USDA in March 1996, asserting that pest surveys results for 1995-96 showed higher levels of host-specific and fruit fly infestations in Mexican orchards than had previously been reported and that there had been procedural irregularities in the rulemaking process that involved violation of federal conflict-of-interest law.[8] The CAC petition argued that the new pest survey results and procedure irregularities invalidated the rulemaking process and requested another public comment period before a final ruling was made to allow avocado imports from Mexico.

2.3 Initial Economic Assessment

USDA's regulatory procedures for SPS decisions require sequential analysis—first determination that there is essentially no risk associated with a proposed rule and second, on that basis, that economic impacts of the rule be assessed. Such a sequential approach to decision making places greater emphasis on risk assessment than on comprehensive cost-benefit analysis. When the mandate of regulatory authorities is stated in such strong terms for protecting the domestic economy from negative SPS externalities arising from trade, as it often is, then product bans and other severe quarantine measures emerge quite naturally as policy outcomes. A product ban is a high level of intervention to address an SPS externality, but a ban does eliminate the externality risk to the extent that legal trade is its proximate cause.

Even within the risk assessment dimension, there is plenty of room for dispute. First, issues arise about whether an externality threat exists in a given situation. Second, a ban may or may not be least trade distorting—

---

[7] For example, *The Washington Post*, 3/11/96, p. A16.

[8] "American Avocado Growers Uncover New Field Surveys on Mexican Avocado Pest Infestations," PR Newswire, 3/28/96.

perhaps there is another way to eliminate the externality risk, one that allows the product to be traded under some specified conditions. Either way, when the policy decision is perceived only in the risk assessment dimension, there is no impetus to ask whether the cost of the policy is warranted by the benefits, that is whether the level of intervention needed to achieve the risk-reduction objective is also desirable on economic criteria, such as maximizing the expected contribution of the affected markets to national welfare.

In the avocado case, the contestation over the proposed rule brought to light information about pest risks that provided the basis for a cost-benefit analysis taking uncertainty about pest infestation into account (Orden and Romano 1996; Orden et al. 2001). The issues that arise in evaluating the economic effects of either full or partial easing of the import ban are illustrated in Figures 1 and 2, assuming a fixed world price for the product and no tariffs or other trade barriers. The first figure shows the effects of free trade when a pest infestation may raise domestic costs. The domestic price P_{D1} falls to the world price P_W and consumer surplus increases (by C+D+E) whether or not an infestation occurs. Producer surplus falls by C+D (the trade effect) and additionally by G (the infestation effect) if pests raises production costs and lower yields with certainty, shifting domestic supply from S to S'. Consumers are always better off, producers are always worse off, and the net effect on welfare (E-G) can be positive or negative. On a probabilistic basis, the expected domestic supply function will lie between S and S', with its location depending on the assumed level of pest infestation risk.

The analysis is more complicated when only a limited quantity of imports is allowed, say due to some technical restriction. Ignoring regional considerations, the limited imports would lower the domestic price if there is no pest infestation, but to P_{D2} in Figure 2 not to the world price level. The effects on consumers, producers and net welfare are fractions of the outcomes with unrestricted free trade. Pest infestation reduces domestic supply and affects the domestic price in the opposite direction from imports. The equilibrium price can rise or fall. When the domestic price rises, as shown from P_{D1} to P_{D3} in Figure 2, consumers are worse off (by c+d). Producers' surplus rises (by c) with the higher prices but falls due to higher production costs (by f+i+k). Producers may be better or worse off than at the initial equilibrium (better if c>f+i+k). Producers may also be better or worse off than with trade but without a pest infestation (better if c+e>i+k). Whatever the outcome for producers, social welfare falls (by d+f+i+k) compared to its level at the initial equilibrium, or (by d+f+i+k+g) compared to its level with trade but without pest infestation.

If the net effect of trade and a pest infestation is for the equilibrium domestic price to fall (not shown in Figure 2), consumers are made better off and producers worse off than without trade or pest infestation. Consumers gain less, and producers may lose more or less than with trade but without pest infestation, and net welfare may rise or fall (compared to the initial equilibrium) depending on whether the net consumer gain from lower prices exceeds the infestation losses of producers.

In their empirical analysis, Orden and Romano and Orden et al. divided the domestic U.S. avocado market into two submarkets—the northeastern winter regional market and the national aggregate for all other regions and seasons. In the northeastern winter regional market, the domestic price was assumed to fall to the price level of exports from Mexico, substantially below the earlier domestic price. For the rest of the U.S., an equilibrium price was determined by domestic supply and aggregate demand with the northeastern winter regional market excluded.

The proposed partial easing of the avocado import ban had expected effects if no pest infestation occurred. In the northeastern region, the winter season price fell by 35% and consumption increased. The domestic price for the remaining aggregated U.S. market fell by 1.3%, as displacement effects from the northeastern winter market were absorbed by a combination of expanded consumption elsewhere and reduced domestic supply. A net national welfare gain of $2.5 million resulted (about 2% of initial total consumer plus producer surplus), mostly due to the lower price in the northeast. Consumer surplus increased by $2.2 million outside of the northeast, but producer surplus fell by a similar amount, so the net welfare gain was small outside of the northeastern winter market. In contrast, a full liberalization of trade (which was not under consideration by USDA at this time) was estimated to depress domestic avocado production by as much as 50% after full adjustment to lower prices, and to raise consumer surplus by nearly $90 million nationwide.

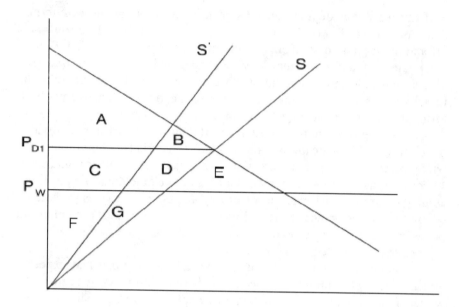

Fig. 1. Effects of free trade with pest infestations affecting supply

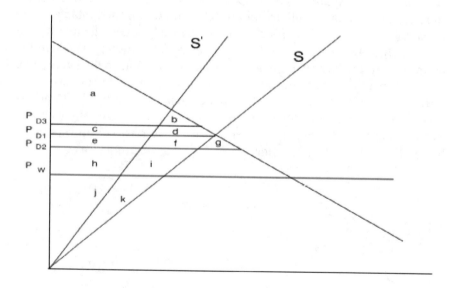

Fig. 2. Effects of limited trade with pest infestations affecting supply

These studies also considered the economic effects of the proposed rule if an avocado pest infestation occurred. A pest infestation increased marginal costs and lowered yields, reducing domestic supply. In the worst-case scenario, reduced availability of avocados under the partial easing of the import ban pushed up the equilibrium domestic price (excluding the northeastern winter regional market) by 30%. The domestic price increase partly offset the effects on producers of lower output and higher production costs but their net loss was $14.7 million, almost seven times as large as from partial easing of the ban alone. A larger economic effect of the pest infestation was felt by consumers outside of the northeastern winter market: their surplus fell by $43.5 million with the increased domestic price. Partial easing of the avocado quarantine would not be sound phytosanitary or economic policy under these circumstances. Yet on a probabilistic basis, it took a much higher likelihood of pest infestation than reported by USDA to turn expected net welfare effects negative. For full trade liberalization, even under the worst-case pest infestation, there was a positive benefit-cost relationship as consumer gains from lower prices more than offset the domestic producer losses.

3   Opening of the U.S. Market

3.1   Partial Easing of the Ban in 1997

Despite continued industry opposition, in February 1997 USDA issued a final rule permitting limited importation of avocados from Mexico under the systems approach. In rejecting the industry arguments about pest risk, USDA reasserted its positive assessment of the safety of the proposed approach and responded to numerous comments received during the public comment period of the rulemaking process. USDA also responded to the concerns raise in the March 1996 CAC petition and subsequent CAC communication about the pending decision. It found neither substantive nor procedural grounds for further delay of a decision to allow limited imports under the systems approach being adopted (USDA 1997). In its economic assessment, USDA evaluated effects of the rule based on diversion of from 10 to 50% of past Mexican exports during November-February to the U.S. market. A diversion of 50% resulted in imports near the level estimated by Orden and Romano. For this level of imports, USDA found similar price effects in the Northeast region and the rest of the country, but its estimates of producer surplus losses and consumer surplus gains were larger. Once the final rule was published, and imports scheduled to be allowed for the first time starting in November 1997, the domestic avocado industry did not file suit to block the USDA decision.

Under the USDA ruling, Mexican avocados began to enter the U.S. market during the winter of 1997-98. After four shipping seasons, no pest infestations had been detected in the imported avocados, lending credibility to the systems approach. Shipments of California avocados to the northeast winter market were largely displaced by imports from Mexico - the California shipments fell to just 1.0 million pounds during 1999-2000 from an average of 7.7 million pounds during 1986-94, as shown in Table 1 (USDA 2001). Wholesale prices of avocados imported from Mexico averaged about 25% less than wholesale prices of domestic avocados during this period. This differential was consistent with the prediction of a regional price difference from the rest of the U.S. market once imports from Mexico became available in the northeast. Avocados from Mexico and California also appear to be imperfect substitutes in the northeast market, where a similar wholesale price differential persisted. Wholesale prices remained above import prices, which averaged about $0.72 per pound. This was consistent with historical import price-wholesale price differentials observed for avocados from Chile (USDA 1997).

Table 1. California avocado shipments (million pounds)

Region	1986-1994 Average			1999-2000 Season		
	Total	Nov-Feb	Nov-April	Total	Nov-Feb	Nov-April
Pacific	128.8	22.8	51.7	150.3	25.0	58.7
Southwest	60.0	14.7	26.7	59.5	11.3	24.9
West Central	12.5	2.8	5.1	15.2	2.9	6.1
East Central	17.6	4.1	7.5	23.1	0.7	5.7
Northeast	16.9	3.6	6.7	24.4	0.3	6.0
Southeast	9.2	2.2	4.0	23.5	4.8	9.7
Total	244.9	50.3	101.8	295.9	45.0	111.2

Source: United States Department of Agriculture, 2001.

With the limited opening of trade under the 1997 rule, imports after the first year averaged over 23 million pounds from over 500 separate shipments (21.5 million pounds in 560 shipments in 1998-99, 25.9 million pounds in 669 shipments in 1999-2000, and 22.5 million pounds in 576 shipments in 2000-01). The level of imports from Mexico were well above the displaced California shipments and nearly double the import demand of 13 million pounds in the Northeast winter market predicted by Orden and Romano at the lower prices expected once imports from Mexico were allowed.

The extent to which Mexican imports exceeded either displacements of California sales or predictions from the economic model suggest that one effect of easing of the quarantine has been expanded consumer demand due to better seasonal availability of avocados. To the extent that market expansion occurs, it provides benefits to consumers and Mexican producers at little cost to domestic producers. Prior to 1997, Chile was the major supplier of avocados during the September-December period, and from 1997 to 2001 Chile accounted for nearly five times as much of the total U.S. supply as Mexico. Avocados from Mexico competed with Chilean exports, but did not dampen total Chilean market sales. The value of avocado imports from Chile grew from $16 million in 1997-98 to $51 million in 1998-99, $35 million in 1999-2000, and $74 million in 2000-01. Simultaneous growth in imports from Mexico and Chile has occurred in the context of a drop in U.S. production, which fell by an average of 35 million pounds during the three seasons 1997-98 to 1999-2000 compared to the average for the two preceding seasons. This shows that imports can serve to stabilize the market in the face of domestic supply variability, thus stabilizing consumer product availability and prices, as well as offering a product competitive with domestic production.

Table 2. Pest risk reductions under a systems approach to importation of Mexican avocados

Risk mitigation measures	Pests of quarantine concern					
	Fruit flies: *Anastrepha spp.*	Small avocado seed weevils: *Conotrachelis spp.*	Avocado stem weevil: *Copturus aguacatae*	Large avocado seed weevil: *Heilipus lauri*	Avocado seed moth: *Stenoma catenifer*	Hitch-hikers and other pests
	Percentage risk reduction					
Field surveys	40 – 60	95 – 99	80 – 95	95 – 99	95 – 99	40 – 75
Trapping and field treatments	55 – 75	0	0	0	0	3 – 20
Field sanitation	75 – 95	15 – 35	70 – 90	15 – 35	15 – 35	20 – 40
Host resistance	95 – 99.9	0	0	0	0	0
Post-harvest safeguards	60 – 90	0	0	0	0	40 – 60
Packinghouse inspection and fruit cutting	25 – 40	50 – 75	40 – 60	50 – 75	50 – 75	30 – 50
Port-of-arrival inspection	50 – 70	50 – 70	50 – 70	50 – 75	50 – 75	60 – 80
Winter shipping only	60 – 90	0	0	0	0	50 – 75
Limited U.S. distribution	95 – 99	95 – 99	90 – 99	95 – 99	95 – 99	75 - 95

Source: United States Department of Agriculture, 2001

3.2   Increased Access in 2001

Based on the early success of the avocado import program, in September 1999 Mexico requested that USDA expand its geographic and seasonal access to the U.S. market. USDA acted within a year to obtain public comments on this request. In November 2001, it issued an amended final rule (USDA 2001). This rule confirmed the risk-reducing effects of the systems approach (see Table 2). The revised rule added access for avocados from

Mexico to a west-central region and increased the shipping season to six winter months. Adding the west-central region increased the domestic shipments with which Mexican avocados would compete from a past average of 7.7 million pounds over 1986-94 to 10.5 million pounds. Increasing the length of the import season increased the domestic shipments with which the Mexican avocados would compete from 7.7 million pounds to 14.1 million pounds for the original access area, and to 19.3 million pounds for the expanded area. Thus, the market access was increased substantially for Mexico by the 2001 rule. Issuance of the revised rule encountered less industry opposition than the initial easing of the quarantine. Still, USDA had to overrule a late CAC petition to suspend its decision process based on a court ruling against the U.S. government on an earlier decision to permit citrus imports from Argentina and the CAC filed suit (still pending) to overturn the new rule.

3.3   Further Opening in 2005

With the additional opening of the U.S. market, avocado imports from Mexico rose from 27.9 million pounds in 2001, to 58.8 million pounds in 2002, and 76.8 million pounds in 2003. The government of Mexico requested in November 2000 that the regulations be amended again to allow importation into all 50 states throughout the year. APHIS undertook another pest risk assessment. Although substantial reductions in risk had been associated with the seasonal and geographic shipping restrictions (see Table 2), APHIS eventually concluded that removing these restrictions while retaining other aspects of the systems approach to risk management would result in fewer than 450 infected fruit entering the U.S. annually, and posed "an overall low likelihood of pest introduction" (USDA 2004). In part this pest risk assessment rested on the six years of accumulated evidence, in which no pests had been detected in over 10 million inspected fruit. New scientific evidence was also available by 2003 demonstrating that the Hass avocado was not a host to certain fruit flies (Aluja et al. 2004). APHIS issued a new final rule on November 30, 2004 that specified conditions for year-around importation of Mexican avocados into 47 states (all except California, Florida and Hawaii) starting in 2005, with access to all states after a two-year implementation delay. Thus, nearly fifteen years after the avocado trade issue was brought to the fore during the NAFTA negotiations, and nearly eight years after the initial partial opening of the

U.S. market, a fundamental reversal of the 1914 ban was accomplished.[9] In doing so, APHIS continued to restrict imports to eligible orchards operating under a systems approach to risk management. Requirements remained in effect for surveys for avocado-specific pests, certification of compliance with pre-harvest and post-harvest handling requirements, traceability, and sample fruit testing. APHIS also continued to require surveying for fruit flies, rejecting the conclusion that Hass avocados were a "non host" in favor of the more conservative status of "very poor host" (USDA 2004).

Projected economic effects of the 2004 final rule are presented in Tables 3 and 4 (USDA 2004).[10] The economic model used for these projections updates average data to a recent two-year base period (October 2001-October 2003) and is more sophisticated than previous modeling in several respects (USDA 2004; Peterson et al. 2004). On the supply side, California, Mexico and Chile are included as producing regions. The year is divided into two periods: October 15-April 15 (period 1) corresponding to the period in which Mexican avocados have been imported under the 2001 rule, and April 16-October 14 (period 2) during which imports from Mexico have not previously been allowed. Avocados from the three countries are treated as imperfect substitutes by consumers, instead of perfect substitutes, accommodating differences in wholesale prices that have persisted by country of origin during the past six years. The Mexican producer price for exported avocados is held constant (at $0.63 per pound) because of extensive additional productive capacity eligible for certification, while supply from California and Chile are price responsive. The fuller specification of the seasonality, substitutability and third supplier allows more precise estimation of the effects of a change in the import rule than would be possible with a simpler model structure such as utilized by Romano and Orden or the earlier USDA assessments. Sensitivity analysis was conducted by simulating the model while drawing its key parameters from assumed random distributions around the benchmark values.

---

[9] Just as the NAFTA negotiations gave a boost to efforts to have the avocado ban reconsidered, intensive discussions between Mexico and the U.S. about bilateral SPS trade regulations after a case of BSE was discovered in Washington state may have created an environment conducive to bringing closure to the assessment of a revised rule on avocados in 2004.

[10] Peterson served as a consultant to USDA in developing the model used for their economic assessment, which is based on earlier model development in Peterson et al. (2004).

Table 3. Estimated near-term changes in annual quantities and prices with 2004 rule

	Initial Prices and Quantities	Importation Excluding CA, FL and HI	Importation into All 50 States
		million pounds	
Quantity total supplied by:	581.071	633.542	660.868
California	346.011	320.821	303.866
Chile	176.814	158.695	147.695
Mexico	58.247	154.026	209.307
		dollars per pound[a]	
Wholesale Price of: Avocados supplied by:			
California	$1.63	$1.43	$1.29
Chile	$1.29	$1.20	$1.15
Producer Price for:			
California	$1.02	$0.81	$0.67
Chile	$0.59	$0.49	$0.44

[a] Prices weighted by regional and time period quantities. Producer and wholesale prices for avocados from Mexico are assumed constant in the model.
Source: USDA, 2004.

The net effect of allowing Mexican avocados into all 50 states year-round is that exports from Mexico increase by 151.1 million pounds (259.4 %), as shown in Table 3, while supply from California falls by 42.1 million pounds (12%) and imports from Chile decrease by 29.1 million pounds (16.4%). Wholesale and producer prices of California avocados fall $0.35 on average over the year (20.8 and 33.3%, respectively), while these prices fall $0.15 for Chile (10.8 and 25.4%, respectively). Consumer surplus rises by $184.4 million within the US, as shown in Table 3, while producer surplus falls by $114.4 million for California, leaving a net U.S. welfare gain of $70.1 million (counting the producer surplus loss of $24.3 million for Chile leaves a net global gain of $45.8 million).

Table 4. Estimated near-term welfare gains and losses with 2004 rule

	Importation Excluding CA, FL and HI		Importation into All 50 States	
	million dollars			
	Change in Welfare[a]	Standard Deviation[b]	Change in Welfare[a]	Standard Deviation[b]
Losses in Producer Welfare				
California	-$71.37	$14.27	-$114.39	$20.48
Chile	-$15.71	$5.29	-$24.35	$5.79
Gains in Consumer Welfare				
Period 1[c]				
Region A[d]	$4.02	$0.99	$7.84	$1.18
Region B[e]	$21.92	$2.08	$29.66	$2.34
Region C[f]	$14.17	$3.34	$27.33	$2.48
Period 2[g]				
Region A	$24.998	$2.70	$32.42	$4.22
Region B	$31.76	$3.38	$41.08	$5.29
Region C	$24.81	$5.29	$46.12	$6.34
Total	$121.66	$3.61	$184.45	$1.93
Net U.S. Welfare Gain[h]	$50.29	$14.27	$70.06	$20.48

[a] The difference between baseline values for October 15, 2001-October 15, 2003 and values with the 2004 rule.
[b] Standard deviations of the sensitivity analysis distributions.
[c] October 15-April 15.
[d] The 31 northeast and central states (and the District of Columbia) approved to receive Hass avocado imports from Mexico during the six-month period October 15-April 15 under the 2001 rule.
[e] Fifteen Pacific and southern states excluding California, Florida and Hawaii.
[f] California, Florida and Hawaii.
[g] April 16-October 14.
[h] The sum of welfare losses for California producers and U.S. consumer welfare gains for all regions and both periods.

Based on the risk assessment, adopting the 2004 final rule to open the U.S. avocado market is consistent with its obligations under the WTO to utilize least-trade distorting SPS measures. In doing so, USDA regulators have been willing to accept a substantial net loss to domestic producers. Peterson et al. show that these losses may be offset over a five year period as avocado demand increases due to population and income growth. But this offset was not incorporated in USDA's analysis, which presented the trade, production, consumption and welfare gains and losses shown in Tables 3 and 4 as the consequences of the 2004 rule.

4   Conclusion

The sequential issuance of the 1997, 2001 and 2004 USDA rules allowing avocado imports from Mexico are an example of successful adoption of a systems approach to risk mitigation. The 1997 rule only opened the market to a small extent, but it did so despite significant domestic industry opposition. The 2001 ruling more than doubled the proportion of the total U.S. market to which Mexico had access, but that proportion remained less than 10%. Economic consequences for the domestic industry, and gains for Mexican producers and U.S. consumers, were relatively limited.

Substantial further progress occurred in 2004 under the precedent set in the first two rules. USDA's initial systems approach rested on numerous risk mitigation measures. Among these, the seasonal restriction of winter shipping only and the limited geographic access, first to 19 then to 34 states, were determined to be necessary components of risk management. Nevertheless, after inspections failed to detect any pest infestations in imports under the system approach, and as scientific evidence became available to substantiate the poor host status of avocados for fruit flies, USDA reconsidered its position and relaxed these two restrictive measures. Net economic effects of this revision to its import rules are much larger than before. Several of the system approach requirements still in place remain subject to question and there may be additional modifications to the required procedures. Either way, the long avocado case from 1991 to 2005 illustrates how difficult it is to make progress on trade expansion when there are complex risk issues at stake and a strong domestic industry is affected by the decision making outcome. It also represents a noteworthy success in this regard.

References

Aluja M, Diaz-Fleisher F, Arredondo J (2004) Non-host status of persea ameri-
     cana "hass" to anastrepha ludens, anastrepha oblique, anastrepha serpenttine
     and anastrepha striata (diptera tiphritidae) in Mexico. Journal of Economic
     Entomology 97:293-309
Josling T, Roberts D, Orden D (2004) Food regulation and trade: toward a safe
     and open global system. Institute for International Economics, Washington
     DC
Orden D, Romano E (1996) The avocado dispute and other technical barriers to
     agricultural trade under NAFTA. Invited paper presented at the conference on
     NAFTA and Agriculture: is the experiment working? San Antonio, Texas,
     November
Orden D, Narrod C, Glauber J (2001) Least trade-restrictive SPS policies: an
     analytic framework is there but questions remain. In: Anderson K, McRae C,
     Wilson D (eds) The economics of quarantine and the SPS agreement. Centre
     for International Economic Studies, Adelaide
Peterson E, Evangelou P, Orden D, Bakshi N (2004) An economic assessment of
     removing the partial us import ban on fresh hass avocados. Selected paper
     presented at the annual meeting of the American Agricultural Economics
     Association, August
Roberts D, Orden D (1996) Determinants of technical barriers to trade: the case of
     US phytosanitary restrictions on Mexican avocados 1972-1995. In: Orden D,
     Roberts D (eds) Understanding technical barriers to agricultural trade. Inter-
     national Agricultural Trade Research Consortium, University of Minnesota,
     Department of Applied Economics, St Paul
USDA (1995) Notice of proposed rule on importation of hass avocados. United
     States Department of Agriculture, Federal Register 7 CFR Part 319, Docket
     94-116-3, pp 34832-34842, July 3, Washington DC
USDA (1997) Importation of hass avocado fruit grown in Michoacan, Mexico.
     United States Department of Agriculture, Federal Register 7 CFR Part 319,
     Docket 94-116-5, February 5, Washington DC
USDA (2001) Mexican hass avocado import program: final rule. United States
     Department of Agriculture, Federal Register 7 CFR Part 319, Docket 00-003-
     4, pp 55530-55552, November 1, Washington DC
USDA (2004) Mexican avocado import program: final rule. United States
     Department of Agriculture, Federal Register 7 CFR Part 319, Docket 03-022-
     5, pp 69748-69774, November 30, Washington DC
USDA/ APHIS (2001) Regulatory impact and regulatory flexibility analysis: the
     potential economic impact of expanded importation of hass avocados from
     Mexico. United States Department of Agriculture/ Animal and Plant Health
     Inspection Service, Washington DC
WTO (1994) The results of the Uruguay round of multilateral trade negotiations:
     the legal texts. GATT Secretariat/ WTO, Geneva

The Labels in Agriculture, Their Impact on Trade and the Scope for International Policy Action

Stéphan Marette

1 Introduction

Both economic growth and increased international trade have put on the shelves many new products, requiring a better mastering of food quality and safety. As incomes rise, consumers are more prepared to pay for quality, and demands for information including labeling and traceability at the world level have gained momentum in many countries. The need for a signal may be even more important when consumers cannot be certain of a product's origin, which is the case when agricultural products from a variety of processors and countries are sold at the retail level with no brand designation.

Today's consumers are faced with a plethora of products certification labels concerning safety, nutrition, geographic origin, organic status, respect of the environment, ethical conditions or fair trade. While a private (manufacturer/retailer) brand belongs to a single firm, labels are used by several producers/firms complying with the label rules. This chapter will focus on these labels and their links with international trade.

The links between labeling and trade are difficult to measure. The availability of data is usually the limiting factor in estimating demand curves or elasticities for specific quality segments. With official statistics (such as Comext by Eurostat or UNCTAD-TRAINS), series of prices and quantities for products are very often aggregated without considering quality differences. Precise data are missing for evaluating the international trade impact coming from labels.

Even though few precise estimates exist, and even though the figures that the various countries put forward are always arguable, some studies (Johnson 1997; Ndayisenga and Kinsey 1994) show that national product quality regulations have a significant effect on agro-food trade. Replicating such studies for the labels regulation would be very hard, since there is a great diversity of labels in each country, and each label concerns a relatively tiny segment of the market (not detailed by the official statistics).

Despite the lack of information regarding the trade issues, this chapter provides clues for thinking about the labeling impact on trade. Before detailing some issues regarding the relationships between labeling and international trade, the paper recalls some effects coming from labeling. For each issue, we present a survey of main contributions in both the empirical and theoretical literatures.

2   A Brief Review of the Main Features Concerning the Label

In agricultural markets, labeling, branding, and/or regulation all serve to mitigate potential inefficiencies resulting from imperfect information about product characteristics. If consumers are not fully informed about product characteristics, they may consume a product with an undesired characteristic or pay a price that does not reflect the quality associated with the product in question. Although a label, a brand, and/or a regulation are proposed as tools for mitigating market failures that have resulted from imperfect information (Akerlof 1970), the instruments themselves may generate other distortions, including antitrust concerns or consumers' misunderstanding.

The agribusiness sector is characterized by the coexistence of multinational companies wielding oligopolistic/oligopsonistic power and farmers with very limited ability to influence prices and capture marketing gains. In the United States and Europe, the degree of concentration in agribusiness varies considerably among states and sectors. The strategies of quality promotion differ a lot according to the concentration in different sectors.

Figure 1 illustrates the different types of organization for signaling quality with the number of competitors or sellers involved in one quality signal, when n sellers are identified by consumers in a downstream market. While a private (manufacturer/retailer) brand (or a trademark) belongs to a single firm, voluntary labels are used by several producers/firms. Mandatory labels are imposed on all sellers. Regarding the labels, Figure 2.1 distinguishes between a geographical indication (GI) and a common label (with, in general, a larger number of sellers, $m'>m$) for insisting on the level of exclusion. A geographical indication excludes the sellers who do not produce in the restricted area, which can be a tool for controlling supply (implying some antitrust concerns). Common labels are used by several producers/firms complying with the label rules and/or having a common characteristic (organic status, respect of the environment, ethical conditions or fair trade) that is not particular to one product.

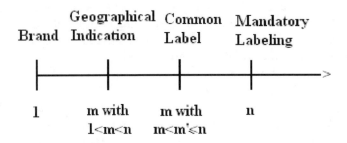

Source: Marette 2005

Fig. 1. The number of competitors involved in one quality signal

Note that there is a great diversity of situations since (*i*) one or several brands may post a geographical indication or a common label and/or (*ii*) several farmers may contract with a brand for the packaging and labeling of a product. Numerous labels are adopted voluntarily, allowing a firm to choose either to label its product or to promote its own brand. Labels are managed by producers/consumers associations, certification firms or non-governmental organizations (NGOs). The state provides property rights protecttion, laws against false characteristics description and sometimes quality-monitoring assistance. In particular, providing standards and guide-lines may be what the government does best. The biggest obstacle here is the credibility of the government itself, but, if the public deems labels to be important, it is an obstacle that a public agency needs to overcome.

Clearly, in a very concentrated industry, the quality promotion is mainly based on brand reputation and private strategies of advertising. The agri-business-multinational companies invest a lot in advertising (Sutton 1991). The existence of economies of scales pushes toward concentration among producers/brands since promotion and advertising imply fixed costs. Because a brand is hard to set up for small industries or scattered farmers, collective labels for promoting high-quality products are necessary.[1]

Label proliferation is the main flaw of the collective labels (Lohr 1998). Consumers Union (2005), a US-based consumer advocacy group, lists over 100 eco-labels on its web site. Just a few of the more well-known labels are the German Blue Angel, the Nordic Swan, dozens of organic certifi-cation labels, "Dolphin Safe," "California Clean," "Bird Friendly," "Shade

---

[1]   Producers' cooperation (or collusion) may be necessary to signal quality when the fixed costs of advertising and third-party certification are large (Marette et al. 1999; and Marette and Crespi 2003).

Grown," and Green Seal. Clemens (2005, p. 8) accounts for "approximately 700 geographical indications (excluding wines and spirits) currently registered in the European Union and the continuous stream of applications to register more products." Peri and Gaeta (1999) count more than 400 official appellations in the wine sector in Italy, 450 appellations in France, and 1,397 in the wine sector in Europe.

The label proliferation may create confusions for consumers. Indeed, Loisel and Couvreur (2001) show that even in France such signals of quality are not clear to many consumers. For example, the recognition of quality labels by French consumers is only 43% for Label Rouge (a high-quality seal for poultry, see Westgreen, 1999), 18% for l'Agriculture Biologique (organic food) and only 12% for Appellations d'Origine Contrôlée (the French GI). One major problem is simply the legibility and clarity of a label, especially one showing some official seal. Although Label Rouge is a well-established label, which suggests that reputation matters, the fact that less than half of French consumers recognize it is suggestive of the problems inherent in any label.

3   The Uncertain Effects of Increased International Trade

In general, policy reform is contributing to a gradual deregulation and trade liberalization, but where food quality is concerned, brand, labeling and regulation are important. As tariffs decrease and/or competition is more intense, the signaling becomes more important for preserving commercial niches.

Trade liberalization and international competition lead to new contexts of competition that modify the signaling strategies. As precise data are missing (see the introduction) and effects are hard to predict, some conjectures are useful for understanding market mechanisms. In a context of perfect information, the opening of a domestic market to imports from other countries results in an increase in domestic welfare. In a context of imperfect information, opening a market to foreign competition increases the incentive for the domestic producer to differentiate itself by improving quality and revealing more information. Consumers may also want to get more information about the origin of products and the conditions of production in foreign countries. These effects may lead to the emergence of new brands or labels, leading to a potential increase of labels proliferation. It should be noted that, except for the wine market and the cheese from Parma, very few GI benefit from an international reputation.

However, if the fixed cost for informing and improving quality is high trade liberalization may result in producers' concentration entailing brands and advertising concentration. Shaked and Sutton (1987) showed that the concentration increases as the market size increases (which is the case with trade liberalization). If quality and information are produced with a fixed cost a firm by selecting a relatively high level of quality can potentially drive competitors with lower quality products out of a market. As fixed cost is not passed to consumers via prices, producers may slash prices for eliminating potential rivals. As a result, concentration at the production level will increase and product variety will decrease in market size. A reinterpretation of this previous result could lead to a reduction of the number of producers and brands coming from the development of international trade. Note that trade liberalization leading to concentration could favor the development of private brands rather than common labels.

These two opposite conjectures show the complexity of the markets effects and it is not obvious to know which effect will dominate. In this context, it is useful to confront the previous implications linked to the increased international trade with the following empirical facts.

3.1   The Need for More Information by Consumers

Some consumers are interested in getting more information about the conditions of production in developing countries. Recently, labels for fair trade and fair working conditions in developing countries gained prominence, even if the market share is relatively limited (between 2% and 4% for different products and locations). Table 1 shows a rapid increase in the production volume under the seal provided by Max Havelaar, one leader of fair-trade certification.

Table 1. World volume of production with the Max Havelaar seal (in tons)

	2001	2002	2003
Coffee	14.432	15.779	19.872
Tea	1.085	1.226	1.989
Bananas	29.072	36.641	51.336
Cocoa	1.453	1.656	3.473
Sugar	468	650	1.164
Rice	0	392	545

Source: http://www.maxhavelaar.org

However, some famous brands only offer a small percentage of their production under the fair trade label.[2] In 2004, only 1% of Starbucks coffee was labeled fair trade, leading to criticisms by some activists about this low volume (Linn 2004). Starbucks responded that it is already a large purchaser of fair trade coffee but that there is not enough of that product that meets its standards.

Table 2 exhibits the cost structure of one packet of coffee in France. The final price difference is mainly explained by the farm gate price between both types of coffee, while the costs are similar for other stages presented in Table 2. The "fairness" in this context comes from the difference at the farm gate price equal to 0.39 euros. Such a premium represents 10% of the final price in the supermarket, which is consistent with the literature findings regarding the price premium.

Table 2. Price of a coffee packet in France (250 gr.) and Arabica from South America

Euros	Without Fair Trade Label	Max Havelaar
Farm gate price	0.19	0.58
Middlemen	0.06	-
Cooperative costs	-	0.08
Exportation costs	0.14	0.14
Max Havelaar fee		0.05
Cost of importation and roasting	1.41 à 2.61	1.45 à 2.5
Final price in supermarket	1.8 à 3	2.3 à 3.35

Source: Lecomte 2003.

Large differences in social conditions/standards in the world explain the demand for ethical characteristics by consumers.[3] The definition of "fairness" is relatively tricky to set up. The Achilles' heel of ethical labeling is the lack of a clear definition combined with a "lenient" certification

[2] Recently, eight companies with brands in France signed an agreement with Max Havelaar for offering products made with "fair" cotton (Les Echos, March 4, 2005, p. 18).

[3] Bigot (2002) examined a variety of attribute signals that might exist in a product and showed that, at least for French consumers, the rank in terms of importance was the absence of child labor, followed by the origin of the products, and decent working conditions for workers who produced the product, positive environmental externalities such as the absence of pollution during the production process. He found that 53% of French consumers would pay a premium for ethical characteristics and this premium would only be 5%. Another 44% would pay no such premium.

process. In this context, the regulation is useful for imposing a clear definition for some labels and/or for controlling the certification activity of private middlemen.

The increased international trade leads to a higher consumers' sensitivity regarding the origin of products. Economists have shown that the origin of food products seems to matter – at least for European consumers. Loureiro and McCluskey (2000) show that the label of origin for fresh meat in Spain leads to price premia for medium quality. Scarpa et al. (2005) and Whirthgen (2005) confirm the existence of consumer preferences for territorial origin of production certification and regional food. Stefani et al. (2005) show that, in the case of Italian spelt, a direct impact of the origin on the willingness to pay exists. Roosen et al. (2003) also suggest that consumers place more importance on labels of origin as opposed to private brands for beef, although this study is applied to European consumers facing the mad cow disease, for which regional labels take on a highly significant meaning. Bazoche et al. (2005) show that label information has an effect during an experimental process that compares the consumers' reactions to French and Californian wines.[4]

The previous developments suggest that a significant effect on prices or consumers' willingness to pay exists, even if the price premium is relatively low. As McCluskey and Loureiro (2003, p. 101) mention, "The major generalization we can draw from [the] group of empirical studies on consumer response to food labeling is that consumer must perceive high eating quality in order for the food product to command a premium. This was particularly important for socially responsible and origin-based products." It means that good quality of products is essential for having a premium with a fair trade label.

3.2 A New Context of Competition

The international competition has deeply reshaped the world market. Development of brands and wineries concentration in Australia and Chile are challenging the leadership of the European GI in world markets.

The wine sector in the European Union is based on the GI for medium- and high-quality wines, where grape production is regulated, with a maximum yield allowed per unit of land. This yield system, which is often

---

[4] Note that these results concern European markets. Even if geographical indications are used less often in the US than in Europe, US farmers are also concerned by this tool, for instance with the Arizona Grown label, Idaho Potatoes, Florida Oranges, Vidalia Onions, Wisconsin Real Cheese, and so forth (Hayes and Lence 2002; Hayes et al. 2004; McCluskey and Loureiro 2003).

disconnected from market demand, does not impede excess supply in some areas, as for the Beaujolais area in France in 2005 (Bombaron 2005). The maximum yield imposed on GI may impede farmers to reach the minimum-efficient scale.[5] Some European GI imposed numerous restrict-tions that stifle the search for commercial efficiency. The excess of regulation for linking origin and quality seems problematic (see Zago and Pick 2004, Ribaut 2005). Conversely, the main features of regulations in the United States, Chile, and Australia are the lack of detailed rules, that is, the freedom to experiment with new techniques; the production and marketing of wines according to single varieties of grapes, sometimes associated with the production region; and a very intense use of marketing investments. All of these features appear to be quite relevant in the world market.

Wineries in Australia are much bigger than the ones in Europe. The average vineyard size in France is less than 2 hectares versus 111 hectares in Australia. Four firms are dominating the Australian market, namely, Foster, Southcorp, Hardy, and Orlando Wyndham. The combined production share of the four largest firms in New Zealand is 85%, while the combined production share of the two largest firms in South Africa is 80%.[6] In other words, the wine promotion in Australia, Chile or the US favors the brand advertising, which facilitates the good reputation and the recognition by buyers. The brand is the most visible information for the Australian wines. This trend seems consistent with the theoretical results of Shaked and Sutton (1987), namely a trend towards more concentration of the brands in a context of increase in market size.

Unlike the industry in Australia or Chile, the wine industry in Europe is very fragmented. The opportunities for mergers in Europe are limited by ownership structures with scattered producers, geographic boundaries, and/or product diversity. Indeed, apart from some notable exceptions, e.g., the Champagne (Economist 2003) or Bordeaux regions, the wine industry in Europe is made up of many small firms, which may lack adequate capital for the necessary investments in new technologies and marketing policies.

---

[5]   Benitez et al. (2005) compare the cost structure of GI producers with non-GI producers for the production of French Brie cheese. They exhibit that GI producers face a more costly production technology and do not profit from scale economies.

[6]   Recent international mergers revamped international wine trading (Marsh, 2003a,b). In 2000, Foster merged with Beringer, a Californian wine firm. In 2003, Hardy merged with Constellation Brands, a U.S. company. As Marsh (2003b) puts it, those mergers undermined Europe's dominance of the sector.

The small size of wineries in Europe reinforces the problem of the proliferation of appellations/wineries (Marette and Zago 2003). The large number of GI assures product diversity but certainly increases buyers' confusion (see Consumer Reports 1997). The recognition of quality labels by French consumers is only 12% for Appellations d'Origine Contrôlée, the French GI system (see Loisel and Couvreur 2001). Berthomeau (2002) discusses the difficulty that the various French appellations have had in entering new export markets because of the absence of any clear specification of the label that distinguishes one appellation from another in consumers' minds. The collective reputation of French wines plummeted during the last decade (Conan 2005; Echikson 2005; Ribaut 2005). The inter-professional group of Bordeaux producers (CIVB, Conseil Interprofessionnel des vins de Bordeaux) completely revamped its generic advertising campaign for reaching consumers of different countries in order to restore its collective reputation (Germain 2005).

In addition, in Europe, the GI system needs to be reformed (Giraud-Heraud et al. 2002; Ribaut, 2005). The Champagne appellation is an example in which the combination of famous brands (with large vineyard size and enough capital for advertising) and a prestigious GI matters for consumers ready to pay a large premium (see Combris et al. 2003). An "efficient" combination of brands and GI also characterizes the Napa Valley appellation, which generates a price premium compared to an equivalent-quality bottle with a different appellation (Bombrun and Sumner 2003). A possible solution for improving the European GI system would consist of simplifying the GI rules by associating brands with a production region such as Bordeaux or Chianti. The issue of GI regarding international trade is maybe overstated since the previous example under-scores the fragility of the GI system for wine coming from the recent changes in the world wine market.

4   Which International Policy Action?

Labeling and consumer information policies are often portrayed as preferable alternatives to regulation because they are cheaper for producers, leave the choice to consumers and are less likely to constitute trade barriers (see Beales et al. 1981 and OECD 1999). Mandatory labels may entail trade distortions or impede the entry of producers who cannot comply with the requirements.[7] Ideally, economists and policy makers

---

[7]   See Bureau et al. (1998), Mahé (1997), Nimon and Beghin (1999), and Sheldon (2002).

have argued that regulators should develop trade policy to cap as much as possible any trade distortions coming from a labeling program (Runge and Jackson 1999). The distortions under a mandatory label are generally lower than the ones coming from an import ban or a minimum-quality standard (see Bureau et al. 1998).

There is an inclination for each country to develop its own system of labels. There is a practical and admittedly simple test to help policy makers discern whether mandatory labeling is being used to increase societal welfare or whether it is being used as a trade barrier (Crespi and Marette 2001 and 2003). Essentially, in a country that requires labeling, if the ratio of consumers concerned about one characteristic to indifferent consumers is low, a voluntary label signaling this characteristic is likely to improve welfare. Conversely, if this ratio is high, then a mandatory label may increase welfare in that country. Thus, observations of governments requiring labels when consumers in those countries show little interest in the debate should be closely examined. Moreover, heterogeneity among consumers may lead to different regulations that may increase the labels proliferation at the international level.

The labeling raises the issue of the access to the domestic market for foreign producers who want to compete in the label niche. Product labeling is theoretically covered by the 1979 Technical Barriers to Trade (TBT) Agreement, but in practice a number of problems arise at an international level with regard to transparency, mutual recognition and control, and these problems proliferate as countries impose their own specifications and labels.

4.1 Mutual Recognition or Harmonization

In principle, foreign producers (with enough capital) may adhere to a voluntary label program and benefit from a collective reputation already established by the common label which should favor entry. The compliance cost linked to the label requirement may ruin the foreign incentive to enter a common label program. This last problem is often crucial for producers in developing country.

The compliance cost explains the effort for harmonizing the label system in the European Union (EU). The European Commission wants to impose the standardization of food labels across the EU. "National laws vary, leading to increased costs for producers for packaging and labeling. Streamlining the various laws will bring considerable cost savings for the food industry, explained Günter Verheugen [the EU industry commissioner]."[8] The labels proclaiming Protected Designation of Origin (PDO) and Protected Geographical Indication (PGI) are already defined at the European level (EEC 1992). The harmonization among different labeling systems is difficult to implement since some countries must make their labels rules more stringent while others must make their labels rules more lenient.

In contrast to standardization (or harmonization), mutual recognition is the alternative way to combine labeling diversity and trade development among countries. Mutual recognition of labeling for organically farmed products is sometimes difficult to achieve because countries apply the relevant criteria more or less strictly, or because some countries are considering granting such labels to genetically engineered or irradiated products. For organic products in Europe or in the US, foreign producers may stamp their products with a domestic organic label under different procedures. The article 11 of Regulation 2092/91/EEC in the EU and the US National Organic Program open up the respective organic food market to products from third countries based on the concept of equivalence. Lohr and Krissoff (2001) showed ambiguous effects of these mutual recognition programs in terms of domestic and exporters' welfare for organic products.

With respect to organic foods, definitions vary a lot among countries. What constitutes an "organic food" has been very difficult to define (Browne et al. 2000). The United States Department of Agriculture's new guidelines on organic food certification came after years of discussion with industry groups as to what characteristics could be considered as organic. The new regulations prevent organic producers from using irradiation to decontaminated products, sewage sludge as fertilizer, and genetically modified ingredients, although some had argued that these techniques did not affect "organic" production since the foods were not produced using conventional chemical fertilizers or pesticides. It is not certain that such a definition is "universal" or applied by other countries or by other private eco-labels. In this debate, the stumbling block is the importance of production conditions for consumers with preferences that vary a lot among countries, impeding the labels harmonization.

The mutual recognition of geographical indications is allowed by the 1994 WTO Agreement on Trade-Related Aspects of Intellectual Property Rights (TRIPS). Geographical indications signaling a particular quality coming from one area are protected under articles 22 to 24 of the TRIPS agreement. If a quality dimension is recognized for a product coming from a single area, no producer external to this area is allowed to mimic the indication. An additional protection for Geographical Indications is provi-

---

[8] See *World Food Law*, February 2005, 80, p.10.

ded for wine (article 23). However, an appellation deemed as "generic" cannot benefit from the exclusive geographical indication (article 24). Controversy arises when names that are protected in one region have a common usage in another. Thus, the term Parmesan protected in Europe is a generic name in the US. The decision concerning the "generic" dimension is decided by national courts. This explains why the name *Chablis* is considered (1) as a generic name that every farmer may use in the US and (2) as a protected geographical indication limited to restricted area of Burgundy in France.

The controversies about the definitions of geographical indications between Europe and the United States (Babcock and Clemens 2004) led to a recent panel on geographical indications (WTO 2005). The panel suggested that some points of the EC regulation 2081/92 regarding the role of governments has to be amended (EEC 1992). In particular, the rights of US trademarks could not be limited by GI regulations. However, the panel recognizes that some articles of the TRIPS Agreement were not violated by the EC regulation 2081/92 (for details see Clemens 2005). A recent agreement between the US and the EU seals mutual recognition of practices for the wine market (USTR 2005). The agreement cancels numerous exemptions that allowed US wine to be imported into the EU. Both countries mutually recognize oenological practices. The US agreed to limit the use of traditional names like Champagne and Chianti which means that they are ready to improve the compliance of some appellations with the requirement of the article 23.

This 2005 agreement on wine between the US and the EU is bilateral. One complementary possibility would be to search for multilateral agreement for the initial definition of the label or the harmonization of labels.

4.2 Labels Defined at the International Level

In a context of labels/appellations proliferation, an international reputation is very hard to acquire because of buyers' confusion and insufficient promotional efforts or education. The small market share of each label does not lead to sufficient economies of scale, since promotion mainly generates fixed costs. One possibility would consist of defining official signs of quality at an international level to reduce label proliferation and possible trade distortions.

The definition of international standards could be organized by forums or by NGOs. This is for instance the case for the fair trade definition. For determining an international standard on what is fair, several national organizations (including Max Havellaar introduced in Table 3) joined the Fairtrade Labeling Organization (FLO 2005).

Few labels defined by international organizations already exist. The MSC label signals sustainable and environmentally responsible fisheries. This label is managed by the Marine Stewardship Council, an independent organization. The Forest Stewardship Council (FSC) delivers the FSC label that signals sustainable developments in the forest management.[9] This international label is a first step in the effort to reduce barriers to certification in developing countries. This label (with 23% of market share for the certified wood in 2002) competes with Sustainable Forestry Initiative (SFI) label in the US (17% of market share) and the label of the Pan European Forest Certification System (PECF) in Europe (38% of market share). Indeed, the FSC label is used by wood producers in numerous countries (see Table 3).

Table 3. Certified forest sites endorsed by FSC in 2004

Continent	Europe	North America	Latin America	Africa	Asia-Pacific	Total
Area certified (million ha)	27.3	9.7	6.4	1.94	1.59	46.9
%	58.1	20.6	13.6	4.1	3.3	100

Source: www.certified-forests.org (accessed in April 2005)

The FSC certification concerns production sites with an average size equal to around 68,500 hectares per site. The increase of the total number of hectares certified with the FSC label over the last decade suggests a

---

[9]   The ISO 9000 certification is also a signal with a world dimension. The focus of ISO is on system quality rather than the quality of the end product, thus ISO 9000 certification in no way ensures that a firm produces high-quality products. This last point explains why we abstract from ISO considerations for the rest of the paper. The International Standards Organization (ISO) based in Geneva, develops "standards" which represent voluntary principles of good practice and the ISO 9000 series of standards detail internationally accepted procedures and guidelines to maintain a consistent quality in product design, production, installation and servicing, and practices for certification. ISO certification then involves a third party certifying that these aspects of a firm's quality management system are in accordance with the principles laid down by the standard. These standards are not intended to replace product safety or other regulatory requirements, but specify those elements that quality management systems must have to produce final products that consistently meet the required specification.

viable existence of a label adopted and recognized in numerous countries.[10] Fisher et al. (2005) note that the standardization of certification programs is unlikely to overcome all the barriers deriving from various certification programs across countries. The effects of harmonized standards for reducing producers' compliance cost could be significant in a sector such as the wood industry.

5   Conclusion

This paper introduced some economic effects linked to labels in a context of international trade. All the results reviewed here suggest that labels often matter to consumers in a context of international trade development. More particularly, the fair trade labels and the identification of origins with GI may lead to a significant premium for producers in developing countries. However, more details and new studies would be necessary for refining the analysis. In particular, the collection of more precise data regarding the market segmentation would be valuable for the analysis.

Eventually, the clarity of the information and the absence of confusion for consumers should guide the private and regulatory intervention at the international level. The main drawbacks are the labels' proliferation and consumers' confusion, which limit the efficiency of such a collective system for signaling quality compared to brands. Clearly, conditions for the success of collective-quality promotion are the absence of signal proliferation and the absence of excess regulation that may create barriers to certification and impede the product differentiation. International trade raises the issue of the mutual recognition versus the standardization of the existing labels among various countries. One possibility for avoiding label proliferation would consist of defining official signs of quality at an international level to reduce label proliferation and possible trade distortions. The definition of international standards could be organized by NGOs.

---

[10] Part of the European furniture industry has signed a charter and is contemplating using only wood that has been granted the FSC or PEFC environmental label.

References

Akerlof G (1970) The market for lemons: qualitative uncertainty and the market mechanism. Quarterly Journal of Economics 84:488-500

Babcock B, Clemens R (2004) Geographical indications and property rights: protecting value-added agricultural products. MATRIC Briefing Paper 04-MBP 7. Midwest Agribusiness Trade Research and Information Center, Iowa State University, May 2004

Bazoche P, Combris P, Giraud-Heraud E (2005) Willingness to pay for appellation of origin in the world chardonnay's war: an experimental study. Mimeo, Institute National de la recherche agronomique, Ivry

Beales H, Craswell R, Salop S (1981) The efficient regulation of consumer information. Journal of Law and Economics 24:491-544

Benitez D, Bouamra-Mechemache Z, Chaaban J (2005) Public labeling revisited: the role of technological constraints under protected designation of origin regulation. Presented at the XIth European Association of Agricultural Economists Congress, Copenhagen, August 24-27

Berthomeau J (2002) Comment mieux positionner les vins Français sur les marchés d'exportation? Ministère de l'Agriculture, Paris

Bigot R (2002) La consommation 'engagée': mode passagère ou nouvelle tendance de la consommation? Les 4 pages des statistiques industrielles, SESSI, DIGITIP. Ministère de l'Economie et des Finances et de l'Industrie, Paris

Bombaron E (2005) Le torchon brûle en beaujolais. Le Figaro, August 12:4

Bombrun H, Sumner DA (2003). What determines the price of wine? The value of grape characteristics and wine quality assessments. AIC Issues Brief 18, Agricultural Issues Center, University of California, Los Angeles

Browne A, Harris P, Hofny-Collins A, Pasiecznik N, Wallace RY (2000) Organic production and ethical trade: definition, practice and links. Food Policy 25:69-89

Bureau JC, Marette S, Schiavina A (1998) Non-tariff trade barriers and consumers' information: the case of the EU-US trade dispute over beef. European Review of Agricultural Economics 25:437-462

Clemens R (2005) Geographical indications, the WTO and Iowa-80 beef. Iowa Ag Review: 8-9

Combris P, Lange C, Issanchou S (2003) Assessing the effect of information on the reservation price for champagne: what are consumers actually paying for? In: Ashenfelter O, Ginsburgh V (eds) The economics of wine. Princeton University Press, Princeton

Conan E (2005) Le prix de l'excellence. L'Express, September 5:48

Consumer Reports (1997). Wine without fuss. 10-16.

Consumers Union. http://www.eco-label.org/labelIndex.cfm. Accessed December 20, 2005.

Crespi JM, Marette S (2001) Politique de label et commerce international. Revue Economique 52:665-672

Crespi JM, Marette S (2003) 'Does contain' vs. 'does not contain': does it matter which GMO label is used? European Journal of Law and Economics 16:327-344

Echikson W (2005) In Bordeaux, the price may not be right. Wall Street Journal, September: W5

Economist (1999) The Globe in a Glass: A Survey of Wine, December 18th-30th: 97-115

Economist (2003), Blended, January 25th 2003, p 61

EEC (1992). Directives n. 2081/92 and n.2082/92 (L. 208), Brussels, Belgium.

Enneking U (2004) Willingness-to-pay for safety improvements in the German meat sector: the case of the Q&S label. European Review of Agricultural Economics 31:205-223

Finke MS (2000) Did the nutrition labeling and education act affect food choice in the United States? Presented at 'The American Consumer and the Changing Structure of the Food System', Economic Research Service, USDA, May 4-5, 2000, Arlington

Fisher C, Aguilar F, Jawahar P, Sedjo R (2005) Forest certification: toward common standards? Resources for the Future 05-10, Washington DC

FLO (2005). Fairtrade Labelling Organization. Website available December 15, 2005. http://www.fairtrade.net/.

Germain S (2005) Le tournant stratégique du vin français. Les Echos, June 21:8

Giraud-Heraud E, Soler LG, Tanguy H (2002) Concurrence internationale dans le secteur viticole: quel avenir au modèle d'appellation d'origine contrôlée? INRA-Sciences Sociales, no. 5-6/01, Ivry

Hayes D, Lence S (2002) A new brand of agriculture: farmer-owned brand reward innovation. Choices, Fall:6-10

Hayes D, Lence S, Stoppa A (2004) Farmer-owned brands? Agribusiness: An International Journal 20:269-285

Johnson R (1997) Technical measures for meat and other products in pacific basin countries. In: Orden D, Roberts D (eds) Understanding technical barriers to agricultural trade. The International Agricultural Trade Research Consortium, St.Paul

Lecomte T (2003) Le pari du commerce equitable. Editions d'Organisation, Paris

Linn A (2004). Starbucks smelling the coffee. Miami Herald, April 17

Lohr L (1998) Welfare effects of eco-label proliferation: too much of a good thing. University of Georgia, Athens

Lohr L, Krissoff B (2001) Consumer effect of harmonizing international standards for trade in organic foods. In: Krissoff B, Bohman M, Caswell JA (eds) Global food trade and consumer demand for quality. Kluwer Academic Publishers, Dordrecht

Loisel JP, Couvreur A (2001) Les Français, la qualité de l'alimentation et l'information. Credoc INC, Paris

Loureiro ML, McCluskey JJ (2000) Assessing consumer response to protected geographical identification labeling. Agribusiness: An International Journal 16:309-320

Mahé L (1997) Environment and quality standards in the WTO: new protectionism in agricultural trade? A European perspective. European Review of Agricultural Economics 24:480-503

Marette S (2005) The collective-quality promotion in the agribusiness sector: an overview, CARD Working Paper 406

Marette S, Crespi JM, Schiavina A (1999) The role of common labeling in a context of asymmetric information. European Review of Agricultural Economics 26:167-178

Marette S, Crespi JM (2003) Can quality certification lead to stable cartel. Review of Industrial Organization 23:43-64

Marette S, Zago A (2003) Advertising, collective action and labelling in the European wine markets. Journal of Food Distribution Research 34:117-126

Marsh V (2003a) BRL hardy soars after constellation talks. Financial Times, January 14

Marsh V (2003b) Australia and US put case for new wine order. Financial Times, January 15

McCluskey J, Loureiro M (2003) Consumer preferences and willingness-to-pay for food labeling: a discussion of empirical studies.' Journal of Food Distribution Research 34:95-102

Ndayisenga F, Kinsey J (1994) The structure of non-tariff trade measures on agricultural products in high-income countries. Agribusiness: An International Journal 10:275-92

Nimon W, Beghin J (1999) Ecolabels and international trade in the textile and apparel market. American Journal of Agricultural Economics 81:1078-1083

OECD (1999) Food safety and quality issues: trade considerations. Consultants' report by J.C. Bureau, E. Gozlan, S. Marette. Organisation of Economic Co-operation and Development, Paris

Peri C, Gaeta D (1999) Designations of origins and industry certifications as means of valorizing agricultural food products. In: Peri C, Gaeta D (eds) The European agro-food system and the challenge of global competition, Ismea, Milan

Ribaut JC (2005) Peut-on encore garantir la qualité? Le Monde, June 17:23

Roosen J, Lusk JL, Fox JA (2003) Consumer demand for and attitudes toward alternative beef labeling strategies in France, Germany, and the UK. Agribusiness: An International Journal 19:77-90

Runge C, Jackson L (1999) Labeling, trade and genetically modified organisms (GMOs): a proposed solution. Center for International Food and Agricultural Policy, University of Minnesota, St Paul

Scarpa R, Philippidis G, Spalatro F (2005) Product-country images and preferences heterogeneity for mediterraean food products: a discrete choice framework. Agribusiness: An International Journal 21:329-349

Stefani G, Romano D, Cavicchi A (2005) Size of region of origin and consumer willingness to pay for speciality foods: the case of Italian spelt. Mimeo, University of Florence, Florence

Sheldon I (2002) Regulation of biotechnology: will we ever 'freely' trade GMOs? European Review of Agricultural Economics 29:155-176

Shaked A, Sutton J (1987) Product differentiation and industrial structure. Journal of Industrial Economics 36:131-144

Sutton J (1991) Sunk costs and market structure. MIT Press, Cambridge

Westgren R (1999) Delivering food safety, food quality, and sustainable production practices: the label rouge poultry system in France. American Journal of Agricultural Economics 81:1107-1111

USTR (2005). United States Representative Executive Office of the President. United States and European Community reach agreement on Trade in Wine. September 15, 2005. Office of the United States Trade Representative, Washington DC, www.ustr.gov (accessed December 15, 2005)

Whirthgen A (2005) Consumer, retailer, and producer assessments of product differentiation according to regional origin and process quality. Agribusiness: An International Journal 21:191-211

WTO (2005) Panel reports out on geographical indications disputes. World Trade Organization, Geneva

Zago A, Pick D (2004) Labeling policies in food markets: private incentives, public intervention and welfare. Journal of Agricultural and Resource Economics 29:150-169

Social Standards and Their Impact on Exports: Evidence from the Textiles and Ready-Made Garments Sector in Egypt

Ahmed Farouk Ghoneim and Ulrike Grote

1 Introduction

In the last two decades, trade barriers have changed dramatically in their nature, moving from a transparent tariff to non transparent and vague non tariff barriers. Standards in general, and labor and environmental standards in particular, have been among the most important evolving trade barriers (Anderson 1995; Anderson 1996). Nowadays, a large proportion of international traded goods are subject to standards. For example, about 75% of EU intra trade and 60% of US exports are subject to standards (World Bank 2001). The Trade Policy Review of the European Union in the year 2000 stated that for the future, the market access conditions for exporters of foodstuffs are likely to be affected by the EU's policy of increasingly taking into account food safety issues (see WTO 2000). Standards, whether environmental or social are playing an increasing role in determining the world trade.

This has been a result of a number of developments including the success of the General Agreement of Tariff and Trade (GATT) and the World Trade Organization (WTO) in lowering tariff rates significantly, the shift of the comparative advantage especially in "sensitive sectors" from developed to developing countries, and the strong political muscles gained lately by concerned interest groups including among others environmentalists and labor unions in developed countries (for a similar argument see for example Bhagwati 1995; Lee 1997).

Such developments created a hot debate about the impact of social policies, particularly environmental and labor standards, on trade between developing and developed countries (Berlin and Lang 1993; World Bank 2001). The debate revolves around the legitimacy of social policies and whether they impact trade flows in a negative manner or a positive one. There has been no clear international consensus on the net costs and benefits arising from such regulations on export dynamics and competitiveness in developing countries (Maskus and Wilson 2000). On the one hand, there has been some evidence that the adoption of common standards tend to reduce imports[1] (World Bank 2001) and that lower standards are associated with a higher revealed comparative advantage (Rodrik 1997) or at least increased labor hiring (and hence increased output and exports) in certain sectors where child labor is allowed (Maskus 1997). Some researchers advocate the need to link trade to the compliance with social standards that require harmonization on a multi-lateral level to stop "racing to the bottom" and ensure "a level playing field" among developing countries (see for example, Adamy 1994; Polaski 2003a; Bullard 2001).

On the other hand, most of the research done in this area that has examined the relationship between trade and labor standards has reached the conclusion that imposing more stringent labor standards on developing countries is a wrong action. It is neither likely to cure the ills of the sensitive sectors in developed countries nor raise the social welfare status of workers in the developing world and hence should not be imposed (see for example, Krugman and Lawrence 1993; Eglin 2001; Golub 1997; Maskus 1997). Empirical research in this field has been scarce to a large extent (see for example Maskus and Wilson 2000; van Beers 1998). The paucity of empirical evidence on this issue has been the main driving force behind the initiation of undertaking this study, which gives an overview of the state of research on this topic worldwide and analyze the impact of labor standards and labels, with special emphasis on child labor in Egyptian export sectors.

2 Identification of Social Standards which Impact Exports in Egypt

Social standards including mainly labor standards and regulations imposed by international organizations and/or major trading partners are expected to have an impact on Egyptian exports. Egypt as a developing country is not expected to have the same rules and regulations concerning labor standards as its major trading partners from the West (mainly the EU and the US which together receive about 70% of total Egyptian exports). For example, Egypt was one of the countries that was against the inclusion of non product-related process and production methods (PPM) under the auspices of the TBT agreement, whereas the EU was in favor of it (cited in Tallontire and Blowfield 2000, p.579). Egypt was also against the US

---

[1] For example, an OECD study found that differing standards and technical regulations in various national markets, combined with costs of testing and certifying compliance with those requirements can constitute between 2-10% of the firm's overall production costs (cited in Stephenson 1997, p. 21).

proposal of establishing a working party on labor standards as suggested by the US in the Seattle Ministerial Meeting (Panagariya 2000).

In the field of labor standards, the analysis in this study is confined to the so called "core labor standards" (Table 1). The focus on such "core standards" arises from the consensus among the researchers that the imposition of such standards should not deprive the developing countries from their comparative advantage and that their negative consequences on the welfare of their economies are likely to be negligible (see for example, Golub 1997; Dessing 1997).

Table 1. The ILO's core labor standards conventions

Convention	Title	Year	Ratified by Egypt
Convention 29	Forced Labor Convention	1930	1955
Convention 87	Freedom of Association and Protection of the Right to Organize	1948	1957
Convention 98	Right to Organize and Collective Bargaining	1949	1954
Convention 105	Abolition of Forced Labor Convention	1957	1958
Convention 100	Equal Remuneration	1951	1960
Convention 111	Discrimination (Employment and Occupation)	1958	1960
Convention 138	Minimum Age	1973	1999
Convention 182	Worst Forms of Child Labor	1999	2002

Source: Singh and Zammit (2000), ILO (2000); cited in McCulloch et al. (2002)

Egypt has adhered to all these conventions. Nevertheless, there might be legislative loopholes and/or deficiency in the enforcement mechanisms of such conventions. The most evident example of such non-compliance is the issue of "child labor" which has been evident in a number of economic activities in Egypt whether in the agriculture or manufacturing sectors. Non governmental organizations (NGOs) estimated that there are about 1.5 million children working in Egypt below the age of 15 in different fields, mainly related to agricultural activities (Bureau of Economic and Business Affairs 2001).

The child labor phenomenon received increased attention due to the popularity of the "unfair competition in trade" raised by the labor interest groups in the developed countries and their fear of "social dumping" and "racing to the bottom" especially after the increased globalization of product and factor markets (see for example Anderson 1995;1996). Such phenomenon has not been researched deeply on country and sectoral levels (Stephenson 1997) and has not been proven empirically (for a review see McCulloch et al. 2002; Bhagwati 1995). There has been neither clear cut evidence that adherence to core labor standards is correlated with other measures of economic development nor that such core labor standards develop in a certain direction simultaneously (World Bank 2001). One empirical study undertaken by Kamal, Paul-Majumder and Rahman (1993) has proved that imposing trade sanctions on Bangladesh for usage of child labor in the garments industry had undesirable effects on poverty and did not stop child labor. On the contrary, the dismissed children were forced to join the informal sector with worse conditions[2] (for more details see McCulloch et al. 2002, p.308, for the results of the study see also World Bank 2001).

Nevertheless the voices for "fair labor standards" still dominate the rhetoric of politicians in developed countries where the national welfare is not the main emphasis, but rather they are driven by the interests of certain groups as labor unions or producers (see for example van Beers 1998). In fact, the EU has adopted a Generalized Scheme of Preferences (GSP) in January 2002 that doubles the tariff cuts for developing countries on a range of sensitive products like agricultural products, textiles, ready-made garments and steel, if the EU finds those applicant countries to protect basic worker rights (Polaski 2003b). This is in line with the previous 1984 GSP amendment under which the US government introduced the possibility of refusing to grant a preferential entry for exports of a beneficiary

---

[2] The legal age for employment in the garments industry in Bangladesh is 14, but until 1992, many younger children than this were working in the garments factories. In 1992, the US introduced a bill aiming at banning the import of items produced by children. Under the threat of the bill, the Bangladesh Garments Manufacturers and Exporters Association (BGMEA) announced the elimination of child labor by October 1994. 50,000 children were dismissed. Since the children had been working to earn money to contribute to their own survival their dismissal left them in even worse circumstances than the conditions of their labor. Most of the children were forced into even more dangerous employment— including prostitution— in the informal sector, and many families, dependant on children's income, faced even greater poverty. The US has undertaken a positive step "in terms of trade" by increasing the import quota from Bangladesh, but does this positive trade effect overcome the negative "social effect" of increased poverty. This is the question that needs to be seriously addressed.

country "which has not taken or is not taking steps to afford internationally recognized worker rights to workers in the country".

The Egyptian legislation contains two major separate laws. One deals with the labor code in general (Law 12/2003)[3] in the labor market, and one deals specifically with rights of the child and hence contains provisions on child labor (Law12/1996)[4]. Confining our analysis to the core labor standards that have been aforementioned, it seems apparent that the Egyptian labor laws comply with the ILO core standards.

Law 12/2003 which amends Law No. 537/1981 improves the conditions concerning the freedom of association and protection, the right to organize, and collective bargaining. Nevertheless, it subjects them to a number of complex procedures that might result in their ineffectiveness in reality. For example, in the collective agreements among labor unions a provision states that they should not contain any contradictions with rules and regulations related to the General Law and ethics, without explicitly identifying what kind of contradictions might occur. Another example relates to the right of the workers to strike which is subject to prior approvals by the labor union that must be notified and they cannot strike without such approval.

Issues of minimum age and worst forms of child labor which are in compliance with core standards from the ILO labor conventions, are well settled in the new comprehensive labor law as well as the child law (Law No. 12/1996). However, their problem lies in enforcement. The major trade partners have accused Egypt of allowing child laborers in different sectors and especially in cotton cultivation working for 11 hours per day

---

[3] For example, the right to strike has been changed by allowing workers now to strike under certain conditions and procedures. On the other hand, it was never allowed to fire workers, not even in difficult economic situations or economic downturns; now it is allowed given certain procedures.

[4] Egypt's Child Law was adopted in 1996, following recommendations by Egyptian social scientists and children's rights advocates aimed at bringing the country's domestic legislation into conformity with the Convention on the Rights of the Child. The Child Law prohibits the employment of children below the age of 14, but allows children between 12 and 14 to receive vocational training from employers and to take part in seasonal agricultural work, provided that the work "is not hazardous to their health and growth, and does not interfere with their studies". The law limits the work-day for children to six hours, only four of which may be consecutive, and requires the provision of one or more breaks totaling no less than one hour per day. The law further prohibits children from working during their weekly days off, official holidays, and between the hours of 8 pm and 7 am (Article 66).

for 6 days per week. Until 1996, child labor in Egypt was governed by the labor law which permitted children to work at the age of 12. The inconsistency between the Child Labor Protection Agreement and Egyptian legislation was rectified with the enactment of a law specifically concerning child labor in 1996 (Al Ahram Weekly Online, 9-15 May, 2002, Issue No. 585).

A comprehensive study prepared by the Government's statistical agency in 1988 indicated that 1.309 million children between the ages of 6 and 14 were employed. In November 1999, the Minister of Social Affairs reported that one million children participate in agricultural activities. Governmental studies also indicate that the concentration of working children is higher in rural than urban areas. Nearly 78% of working children are in the agricultural sector. However, children also work as domestic workers, as apprentices in auto repair and craft shops, in heavier industries such as construction, in brick-making and textiles, and as workers in tanneries and carpet-making factories. While local trade unions report that the Ministry of Labor adequately enforces the labor laws in state-owned enterprises, enforcement in the private sector, especially in family-owned enterprises, is lax. Many of these children are abused by their employers and are overworked, and the restrictions in the Child Law have not improved conditions due to lax enforcement on the part of the Government (Bureau of Democracy, Human Rights, and Labor; U.S. Department of State 2000).

Hence, the question arises whether Egypt's adherence to core labor standards could make it easier to utilize its comparative advantage. In other words, does imposing standards, especially those related to child labor as they are the most evident, deprive Egypt from fully utilizing its comparative advantage by lessening the market access of its products in its major trading partners?

3   Empirical Analysis of the Textiles and Ready-Made Garments Industry

The textiles and ready-made garments industry was chosen for further analysis as it is a sector, likely to be affected by social standards. According to the data set of the Federation of the Egyptian Industries, there are more than 3000 firms in this sector. In addition, there are many firms that are not registered by the Federation[5]. Out of this data base, a small sample of 83 firms was surveyed in 2004. It cannot be considered as a representative sample but it gives a first impression about the situation related to child labor in the textiles and ready-made garment sector in Egypt.

Out of the 83 firms 81 were private whereas two were public firms. The sample was geographically distributed among the following five governorates: Greater Cairo consisting of Cairo, Kayobia and Giza (65 firms), Mahalla Kobra (9), and Alexandria (9). This geographical distri-bution reflects the nature of the industry which is rather characterized by clusters concentrated in the three aforementioned governorates. All sur-veyed firms focused on the export business. 49 companies (66%) export more than 50% of their output - in terms of value and volume; only 10 companies (11%) export less than 10% of their output.

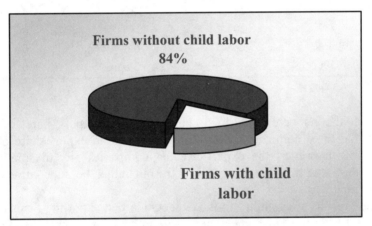

Source: Own survey.

Fig. 1. Distribution of firms according to their use of child labor

The phenomenon of child labor was evident in 13 firms or 16% of the sample (see Figure 1). Five of these 13 firms (6%) have 10 to 20 child laborers below the age of 16; the other 8 firms have less than 10 child laborers. It can be further observed that relatively more girls than boys are employed as child laborers. Child laborers had been employed more often on a temporary basis rather than on a permanent one – there was only one company with 12 permanent child laborers.

---

[5] According to an interview held with the managing director of the Industrial Chamber of Textiles and Ready Made Garments.

It has been also found from the sample that the firms with child labor are all private sector companies. The two public firms do not employ any child laborers. It is also interesting to note that most of the companies with child labor are companies that do not have foreign affiliation; only one subsidiary of a multinational company (out of nine) and one joint venture company (out of six) reported that they were employing child laborers, but both of them only on a temporary basis. The firms with child labor are also mainly small-scale firms as can be seen from the following Table 2.

Table 2. Occurrence of child labor by size of company

Number of workers	Number of firms	In %	Number of firms with child labor
Up to 100	28	33.7	9
100 to 499	27	32.5	4
500 to 1499	20	24.1	0
1500 – 4999	5	6.0	0
More than 5000	3	3,6	0

Source: Own survey.

Child laborers were found in six companies with export shares of more than 50%; three companies with child labor had an export share of less than 10%. However, the export-oriented companies are all small-scale companies, and may not export directly but rather be subcontracted by larger firms.

These results are contrary to the expectation that exporting firms do not hire child laborers. This expectation came in contrast to what we assumed that the level of awareness of labor standards in general and child labor prohibition in specific is rather a common aspect of all exporting firms in the textiles and ready-made garments industry. However, the first evident issue that we arrived at is that the level of awareness is rather low among exporters.

3.1 Destination of Exports

The destinations of exports were mainly the United States and the European Union as shown in Figure 2. However, the exports of firms that hire child labor were mainly directed to Arab countries as shown in Figure 3. This implies that there is a certain level of restrictive measures and a certain level of awareness among exporters on the countries that are less rigid regarding exports that include child labor.

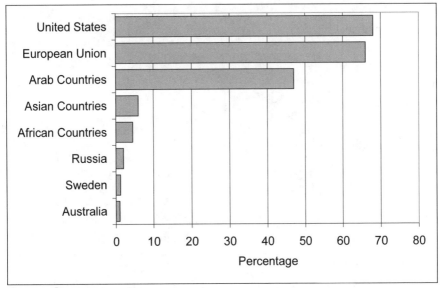

Source: Own survey

Fig. 2. Distribution of firms according to the destination of exports

The main issue that we deduce from this section is that child labor is a phenomenon that exists in the textiles and ready-made garments industry with a relatively high share in the sample (16%). The phenomenon is not correlated with the destination of exports where all destinations receive exports that embody child labor; however, the non Western destinations seem to receive the lion's share. Finally, and contrary to our expectations, a large proportion of firms with child labor exports more than 50% of their output. This implies that child labor is not an impediment to export as usually mentioned in the literature, and it indicates that the level of aware ness among Egyptian exporters regarding this issue is relatively low. One possible explanation of the fact that child labor is not an impediment to export is that some sort of export diversification takes place in the exporting firms; this suggests that firms diversify by additionally exporting to countries with lower standards.

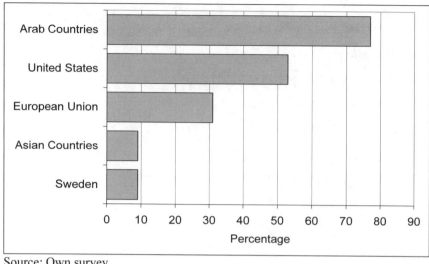

Source: Own survey

Fig. 3. Distribution of firms with child labor according to their export countries

3.2   Awareness about Labor Standards Including Child Labor

The case study also indicates that the level of awareness among Egyptian exporters regarding core labor standards in general and child labor in specific is relatively low. As can be seen from Table 3, many firms hiring child laborers were not aware of the need to comply with labor standards. This is also true for about 50% of the firms without child labor. Table 4 shows the awareness of firms with child labor about individual selected standards. As can be seen, none of the firms declared that they were aware that it was prohibited, whereas they emphasized that they were aware of other core labor standards of ILO, mainly the issues of discrimination and the right to organize and collective bargaining.

Table 3. Are you aware of the need to comply with labor standards?

	Firms with child labor		Firms without child labor	
	Frequency	%	Frequency	%
Aware	4	30.8	34	48.6
Not aware	9	69.2	36	51.4
Total	13	100	70	100

Source: Own survey

Table 4. Degree of awareness of core labor standards in the firms with child labor

	Know		Don't know	
Core labor standards	No.	%	No.	%
Prohibition of child labor	0	0	13	100
Right to Organize and Collective Bargaining	11	85	2	15
Freedom of Association and Protection of the Right to Organize	5	38	8	62
Minimum Age	8	62	5	38
Discrimination (Employment and Occupation)	12	92	1	8
Equal Remuneration	8	62	5	38

Source: Own survey

On the other hand, we find that most firms hiring child laborers believed that imposing labor standards is important to protect the welfare of workers and enhance the market access of their products to the foreign markets. It is interesting to point out that the social objectives dominated the thinking of such firms where they believed that protecting welfare of workers play a predominant role in this regard. Furthermore, 90 to 95% of the firms believe that complying with labor standards will enhance their market access to developed countries and increase the consumer acceptance of imported products. Almost 70% even indicated that they find it important to ensure fair competition. On the other hand, the firms were also asked about the additional costs due to complying with labor standards in general. More than half of the firms estimated the costs of complying with labor standards to be up to 5%. However, many firms also assumed the additional costs to be much higher (Table 5).

Table 5. Additional costs that will or did occur in case of complying with labor standards

Additional cost in percent	No. of firms	% of firms
Up to 5%	47	56
5-10%	23	27
10-15%	6	7
15-20%	3	4
More than 20%	1	1

Source: Own survey

3.3 Child Labor and Schooling

The firms with child laborers were also asked about the percentage of children going to school. Six of them indicated that their child laborers do not go to school, two answered that 10 to 15% of their children go to school, while another five said that 50 up to 80% of their children attend school. Those firms with a higher percentage of school attendance also tend to have more child laborers (10 – 20). This might be explained by the fact that each child works fewer hours because of schooling. We also checked whether firms with an export share of more than 50% and exclusively exporting to the West have a higher percentage of temporary child workers compared to firms with a lower export share and not exclusively exporting to the West. We find that the percentage of temporary workers is almost identical across both types of firms. In addition, five firms indicated that they offer special incentives like free books, bonuses, or even a pay rise to encourage them to go to school. Of the 13 firms, five also said that adults can replace children without adding to the production costs, while eight indicated that additional costs would occur.

3.4 Motivation for Hiring Children

Being asked about the reasons of hiring child laborers, next to 'lower wages', the firms named reasons like 'more skilled', 'preparing well trained skillful workers for future work', 'developing their skills', and 'their financial need' or 'higher learning curve in addition to giving them a better life'.

3.5 Enforcement of Labor Standards

The reason for complying with the prohibition of child labor was mainly based on mutual agreement with the importing country. The role of inspection and monitoring bodies from both local and foreign authorities was considered as being limited (see Figure 4). However, being asked about who should monitor and label labor standards in their field of business in case an agency would be established, about 70% of the firms said that it should be a domestic authority, 22% favored a foreign authority and 7% a joint authority. The major reason for favoring a domestic authority was that they are more aware about the local conditions and that the foreign governments should be prevented from interfering in domestic concerns. On the other hand, the corruption of the domestic system, and the efficiency, commitment and skills were named as reasons for favoring a foreign authority. Almost 80% of the firms said that on-site monitoring by an agency would be needed, while 20% did not like the idea of on-site monitoring.

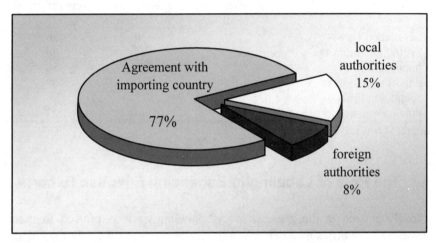

Source: Own survey

Fig. 4. Distribution of firms according to the reason of following a specific requirement in production

Being asked about the role of trade unions, 20 firms answered that they had workers who are members in trade union, and 22 firms (26%) believed that trade unions are able to impose their demands on the entrepreneurs.

As regards measures which can help to decrease child labor, the opinion was divided. While 12 firms said that strict sanctions would help, 17 suggested the establishment of schools and education centers. Another 9 firms asked for subsidies for educating children and 7 generally indicated that the alleviation of the standard of living would help. Interesting results on child labor enforcement are also derived from the following Table 6 which shows that fines and export stops for non complying firms are considered as the most effective ways of reducing child labor. This opinion is shared by 70 to 80% of the firms. Public awareness raising, legislation and laws, and setting a minimum age are, however, seen as no effective measures in reducing child labor.

Table 6. Attitudes on how to reduce child labor

Rank these ways to stop child labor	Not effective / not effective at all	Limited effectiveness	Effective / very effective
Legislation / laws	33	14	36
Setting minimum age	21	26	36
Imposing fines	8	7	56
Export stop for non complying firms	8	7	68
Public awareness raising	46	14	23

Source: Own survey.

4   The Role of Labeling in Enhancing Egyptian Exports

Closely related is the issue of social labeling as it is praised in many developed countries as a very attractive instrument to raise labor standards in developing countries. Social labels provide information via product labels on whether acceptable labor standards were applied in the production process, including the sensitive question of whether child laborers were employed or not.

The origins of social labeling can be traced back to the White Label initiated by a US labor union already in 1899, declaring that clothing had been produced without women and child laborers. Nowadays, social criteria like 'no child labor', freedom of association, wage levels, working hours etc. have been developed for labels especially in the carpet market, the footwear and sports industry but also in the agricultural sector and the textile industry. The labels are known under names like Rugmark, Kaleen, Step, Pro-Child, Care & Fair or Reebok (see Appendix 3.).

In the surveyed textiles and ready-made garments industry in Egypt, labels indicating that no child labor has been involved in the production process are not known to the entrepreneurs. This refers to their own produced products as well as to the inputs they use in their production process. However, two companies state that they have received a certificate that says that they are not using any child laborers. They are being monitored regularly, one of them once and the other one four times a year. For one enterprise, a contract with the importer even specifies that no child labor is allowed in production. 41 companies or 55% indicate that inspecttions of their company took place to control the use of child laborers.

In general, the attitude towards labeling is divided. Out of the total 83 enterprises, 13 or 20% apply positive attributes to labeling. However, 46 enterprises which is around 66%, do not consider labeling a positive thing, and 21 or 30% do not know. 38 or 64% of the companies do not find it important to receive a certificate saying that they use no child laborers. Four companies (6%) find it too costly, while two companies state that such a certificate is a precondition for entering other certification schemes like ISO (Table 7). Most entrepreneurs believe that the awareness regarding labor standards can be enhanced through formal training conducted by domestic authorities, or seminars and workshops, and publications.

Table 7. Reasons for not certifying

Why they don't have this certificate	Frequency	Percent
Not important	46	65.7
Don't know	21	30
Costly	4	5.7
It's a condition/requirement for other certificates (e.g. ISO)	2	2.9

Source: Own survey.

Interestingly, 43 entrepreneurs believe that child labor can be abolished by increased market access. Another 24 indicate that financial aid, reduced taxes or subsidization of raw materials is needed to overcome the cost of abolishing child labor, while 9 entrepreneurs state that no incentives are needed because it is obligatory to abolish child labor, and two indicate that child labor should not be stopped as work helps and protects the children.

Asking about the allocation of responsibilities for implementing labeling schemes, the majority of entrepreneurs (71%) clearly stated that domestic authorities should be responsible for its implementation. Only 22% of the entrepreneurs stated preference for a foreign authority and 7% for a joint authority in charge. The major reasons behind preferring a domestic authority refer to the fact that local entities have a higher awareness about local conditions. In addition, there is a general concern that foreign governments might interfere in domestic matters. Only a few entrepreneurs indicate the issue of corruption of local authorities, and one indicates that foreign authorities understand the export process better than the domestic authorities.

5   Econometric Analysis

The main purpose of this section is to identify which factors related to labor affect the export performance of a firm. More in detail, we want to find out whether labor standards in general and child labor in specific have an impact on the export performance of an enterprise. Export performance is captured as more than 50% of exports of a firm and exclusively to the West. We thus define the dependent variable as the probability of a firm to export more than 50% and exclusively to the West. The independent variables have been grouped into the four categories 'awareness', 'compliance', 'enforcement', in addition to 'firms' characteristics':

$$Prob_i \ (export \ performance) = f(A_i, \ C_i, \ E, \ Z_i) + e_i \qquad (1)$$

where:
A_i = awareness of labor standards in enterprise i,
C_i = compliance with labor standards in enterprise i,
E_i = enforcement of labor standards,
Z_i = firms' characteristics,
e_i = random variable.

For the first three categories, a factor analysis has been conducted to reduce the number of variables by extracting the relevant factors[6]. The variables used are listed in the appendix. The relevant factors which were identified as being relevant are:

• Awareness Factor 1: Opening markets for exports (AwareFAC1)
• Awareness Factor 2: Fairness and awareness (AwareFAC2)
• Awareness Factor 3: Competitiveness through standards (AwareFAC3)

---

[6] For the factor analyses, the test of sphericity with its "measure of sampling adequacy (MSA)" shows to what extent the variables belong together and thus, indicates whether the factor analysis is useful or not. The MSA criterion allows an assessment of the correlation matrix as a whole and also of individual variables. In our calculations, the general MSAs reached values between 0.6 and 0.7 which is assessed as being mediocre; however, the values partly rose to close to 0.7. MSAs > 0.7 are considered as 'middling'. MSAs of individual variables can be assessed from the anti-image matrices (Annex); all variables with MSAs < 0.5 were successively excluded from the calculations. The remaining ones even had MSAs of > 0.8 and > 0.9 which is considered as 'meritorious' and 'marvelous', respectively. For extracting the factors, the Kaiser-criterion was used which means that only factors with eigenvalues of more than 1 are selected.

- Compliance Factor 1: Compliance with selected labor standards (ComplyFAC1)
- Compliance Factor 2: paid vacation and minimum age (ComplyFAC2)
- Enforcement Factor 1: market access & fair competition (EnforceFAC1)
- Enforcement Factor 2: fines, costs & mutual agreement (EnforceFAC2)
- Enforcement Factor 3: control of child labor (EnforceFAC3)
- Enforcement Factor 4: controls of labor standards (EnforceFAC4)

Thereafter, the factors along with a number of firms' characteristics were used as inputs in the logistic regression equation. The firms' characteristics are the following variables:

- Since when has the firm been in business? (firmAGE)
- How many workers do you have? (NoofWorkers)
- Is the firm specialized in its sortiment? (Specialized)
- Do you have a foreign affiliation? (foreignAffilia)
- How many export destinations do you have? (Nodestinations)

Table 8. Logistic regression results for the probability of good export performance

	B	S.E.	Wald	Df	Sig.	Exp(B)
firmAGE	-,018	,032	,307	1	,580	,982
NoofWorkers	,001	,001	4,655	1	,031**	1,001
Specialized	-3,533	2,510	1,981	1	,159	,029
ForeignAffilia	-2,389	2,125	1,264	1	,261	,092
NOdestinations	-3,794	1,379	7,570	1	,006***	,023
EnforceFAC1	1,974	1,124	3,081	1	,079*	7,197
EnforceFAC2	-2,656	1,141	5,416	1	,020**	,070
EnforceFAC3	,031	,670	,002	1	,963	1,032
EnforceFAC4	1,820	1,148	2,513	1	,113	6,171
AwareFAC1	,028	,664	,002	1	,967	1,028
AwareFAC2	-,598	,683	,768	1	,381	,550
AwareFAC3	2,010	1,241	2,623	1	,105	7,460
ComplyFAC1	,386	1,594	,059	1	,809	1,472
ComplyFAC2	-1,490	1,264	1,390	1	,238	,225
Constant	43,686	64,491	,459	1	,498	9,39E+18

Selected cases: 59; -2 Log likelihood: 30,813; Cox & Snell R Square: ,557; Nagelkerke R Square: ,756
*** Significance level 1% ** Significance level 5%; * Significance level 10%
Source: Own calculations.

The results of the logistic regression[7] (Table 8) show that the 'number of workers', 'number of destinations', as well as the first two enforcement factors 'market access and fair competition' and 'fines, costs and mutual agreements' have significant partial effects. This means that the larger the enterprise in terms of 'number of workers', the more likely it is a good export performer identified by a high share of exports and exclusively to the West. This result also supports the descriptive statistics (Table 3) showing that the larger enterprises are less likely to employ children as workers. The significant variable 'number of destinations' has a negative impact on the probability because the more markets (Western and non Western) the firm exports to the more likely it has to comply with different types of standards which negatively affect its export performance. Hence, geographical concentration would enhance the export performance of the firm.

The results also support the hypothesis that firms that are more aware of market access and fair competition considerations are likely to export more than 50% and exclusively to the West. The second enforcement factor "fines, compliance costs and mutual agreements" has a significant negative effect on the probability of having good export performance. This can be explained by the high loading of the compliance cost variable which indicates that the higher the costs are or are expected to be, the less likely the firm is to perform well in terms of exporting more than 50% exclusively to the West. Also the fourth enforcement factor has a positive effect on the probability, though not significant. It indicates that domestic and foreign controls improve the enforcement of labor standards and thus lead to better export performance results. On the other hand, the actual control of a firm (enforcement factor 3) for child labor hardly has any effect on the performance.

In summary, the results show that several variables related to labor standards and child labor have an effect on the probability of a firm to perform well in the export business. In general, it has been found that variables which ensure the enforcement of labor standards have a higher explanatory power for the probability of better export performance than compliance and awareness variables. The aspect of child labor seems to explain the dependent variable to a much lower extent than labor standards in general. Thus, the two factors related directly to child labor standards (ComplyFAC1 und EnforceFAC3) show a highly insignificant effect on

---

[7] The level of predictive power by the regression is quite high. The estimated relationship for the probability to perform well correctly predicts 88% of the observations. Also the R squares indicate that the dependent variable is well explained by the independent variables.

the export performance. For example whether a firm employs child laborers or not has significant effect on whether the firm has good export performance and exclusively to the West.

6   Conclusion and Policy Implications

In general, the study shows that some variables related to labor standards and child labor have an effect on the probability of a firm to perform well in the export business. It has been found that variables which ensure the enforcement of labor standards have a higher explanatory power for the probability of good export performance than compliance and awareness variables. This is along the lines of what the descriptive results show; thus, fines and export stops for non complying firms are considered as the most effective ways for reducing child labor, while public awareness raising campaigns, legislation and laws, and setting a minimum age are seen as rather ineffective measures.

The aspect of child labor seems to explain the dependent variable to a much lower extent than labor standards in general. The two factors which include specifically child labor aspects, namely the enforcement factor 3 (control of child labor) and the compliance factor 1 (selected labor standards), are both highly insignificant. Thus, for example whether a firm employs child laborers or not has not a significant effect on whether the firm exports exclusively to the West. This result is supported by the descriptive analysis showing that child labor is a phenomenon that exists in the textiles and ready-made garments industry with a relatively high share in the sample (16%). All destinations receive exports that embody child labor; however, the non Western destinations seem to receive the lion's share. Moreover, it has been found that a large proportion of firms with child labor export more than 50% of their output, and firms with child labor are mainly small-scale firms.

Econometric analysis shows a strong positive relationship between higher standards and the likelihood of enhanced export performance to the West. However, for those firms with a high volume of exports to Arab countries and for smaller firms (both exporting to the West or Arab countries), the effect of standards might lead to the need of more export diversification within a specific region either to the West or to any other region to build on the economies of scale resulting from harmonized or similar standards.

The results of the case study on child labor in textile and garment enterprises in Egypt revealed that firms with child labor are mainly small-scale firms. Even companies with export shares of more than 50% employ child laborers; however, these exporting firms target their exports mainly to Arab countries, not to the EU or the US. It has been also found that the awareness about the prohibition of employing child laborers is generally relatively low.

It has been also found that labels indicating that no child labor has been involved in the production process are not known to the entrepreneurs in the textiles and ready-made garments industry in Egypt. In general, the attitude towards labeling is divided, however, with the majority of enterprises applying negative attributes to labeling. Interestingly, half of the entrepreneurs believe that child labor can be abolished by increased market access. Nevertheless, the labeling proposal suggested in the study and used intensively by some countries as Bangladesh and India seems to be a reasonable solution. As argued by Freeman (1996) it gives the consumer purchasing a good to freely decide whether he weighs the normal cost of hiring child labor higher or lower than his moral values.

Other policy implications that can help policy makers to overcome the negative impacts of social regulations, if any, is undertaking mutual recognition agreements (MRA)[8] as the study showed that they adhere to labor standards only when the importer requires. MRAs in the field of labor standards can overcome the negative effect of different standards prevailing in Egypt and its major trading partners (despite the fact that evidence has shown that it has not been successful in the EU trials with its trading partners, see Stephenson 1997), taking in consideration the possible scope for harmonization due to the limitations arising from different economic circumstances, and developmental differences (as argued by Anderson 1995).

---

[8] In MRAs, manufacturers are able to obtain required national certificates at the location of production, rather than pay the higher costs of offshore certification. The MRAs are in general applied to technical standards and regulations, and quality management systems. Their application in the field of compliance with labor standards has not been mentioned in the literature according to the knowledge of the authors. Nevertheless, there is nothing that prevents the adoption of such systems in the field of labor standards.

References

Adamy W (1994) International trade and social standards. Intereconomics 29:269-277

Anderson K (1995) The entwining of trade policy with environmental and labour standards. In: Martin W, Winters LA (eds) The Uruguay round and the developing economies. World Bank Discussion Paper No 307, World Bank, Washington DC

Anderson K (1996) Social policy dimensions of economic integration: environmental and labour standards. NBER Working Paper No 5702

Berlin K, Lang JM (1993) Trade and the environment. The Washington Quarterly 16:35-51

Bhagwati J (1995) Trade liberalization and fair trade demands: addressing the environmental and labour standards issues. World Economy 18:745-59

Bhattarcharya D (2002) International trade, social labelling and developing countries: the case of Bangladesh's garments export and use of child labor centre for policy dialogue. Dhaka

Bullard N (2001) Social standards in the international trade. Report prepared for the Deutscher Bundestag Commission of Enquiry Globalization of the World Economy — Challenges and Answers, Thailand, Available on the website: http://focusweborg

Bureau of Economic and Business Affairs, US Department of State (2001) 2000 Country Reports on Economic Policy and Trade Practices, Washington DC

Bureau of Democracy, Human Rights, and Labor - US Department of State (2000) 1999 country reports on human rights practices. February 25, 2000. http://www.usemb.se/human/human1999/toc.html

Dessing M (1997) The social clause and sustainable development. BRIDGES Discussion Papers Vol 1, No 1, International Centre for Trade and Sustainable Development, Geneva

Eglin R (2001) Keeping the t in the WTO: where to next on environment and labor standards? Journal of Economics and Finance 12: 173-191

Ghoneim AF (2000) Antidumping under the GATT and the European Union rules: prospects for the Egyptian European partnership agreement. In: Nassar H, Naeim A (eds) The Egyptian exports and the challenges of the twenty first century. Center of Economic and Financial Research and Studies, Cairo, pp 184-248

Golub SS (1997) International labor standards and international trade. IMF Working Paper No WP/97/37, IMF, Washington DC

Kamal GM, Paul-Majumder P, Rahman MK (1993) Economically active children in Bangladesh. Unpublished report, Associates for Community and Population Research, Dhaka

Krugman P, Lawrence R (1993) Trade, jobs and wages. NBER Working Paper No 4478

Lee E (1997) Globalization and labour standards: a review of issues. International Labour Review 2: 173-189

Maskus KE (1997) Should core labor standards be imposed through international trade policy. World Bank Policy Research Working Paper No 1817, Washington DC: World Bank

Maskus KE, Wilson J (2000) Quantifying the impact of technical barriers to trade: a review of past attempts and the new policy context. Paper presented at the World Bank Workshop on Quantifying the Trade Effect of Standards and Technical Barriers: Is it Possible?, April 27, 2000

McCulloch N, Winters, LA, Cirera X (2002) Trade liberalization and poverty: a handbook. Center for Economic Policy Research/ United Kingdom Department for International Development, London

Panagariya A (2000) Trade-labor link: a post-Seattle analysis mimeo. University of Maryland, http://www1.worldbank.org/wbiep/trade/videoconf/panagariya.pdf

Polaski S (2003a) Trade and labor standards: a strategy for developing countries. Carnegie Endowment for International Peace. http://12.150.189.35/pdf/files/Polaski_Trade_English.pdf

Polaski S (2003b), Labour Standards: Why Happy Workers are Good for Growth, South China Morning Post, 20 January, 2003

Stephenson SM (1997) Standards and conformity assessment as nontariff barriers to trade. World Bank Policy Research Paper No 1826, Washington DC: World Bank

Tallontire A, Blowfield ME (2000) Will the WTO prevent the growth of ethical trade? Implications of potential changes to WTO rules for environmental and social standards in the forest sector. Journal of International Development 12:571-584

Trebilcock JD (1999) Trade labour standards and development: challenges for research from the policy consensus. manuscript

Trebilcock JD (2002) Trade policy and labour standards mimeo. University of Toronto, Toronto

Raynauld A, Vidal JP (1998) Labor standards and international competitiveness: a comparative analysis of developing and developed countries. Edward Elgar Publishing, Massachusetts

Rodrik D (1997) Has globalization gone too far? Institute for International Economics, Washington DC

Van Beers C (1998) Labour standards and trade flows of OECD countries. World Economy 1: 57-73

World Bank (2001) Global economic prospects and the developing countries. World Bank, Washington DC

WTO (2000), European Union trade policy review. World Trade Organization First Press Release, Press/TPBR/137, http://wwww.toorg/english/tratop_e/tpr_e /tp137_ehtml

Appendix: List of Variables Used for the Factor Analysis

1) Awareness of Labor Standards in Enterprise i:

- Do you agree that these labor standards are very important to increase consumer acceptance of imported products? (1=very important; 0=otherwise)
- Do you agree that these labor standards are important to enable access to developed countries' markets? (1= important; 0=otherwise)
- Do you think that labor standards are very important to protect the welfare of your workers? (1=very important; 0=otherwise)
- Do you agree that these labor standards are important to ensure fair competition? (1=very important; 0=otherwise)
- Do you think that public awareness raising will stop child labor? (1=yes; 0=otherwise)
- Do you think labor standards can affect the competitiveness of your exports positively? (1=yes, 2=otherwise)
- Are you aware of the need to comply with labor standards in some countries? (1=yes; 0=otherwise)

2) Compliance with Labor Standards in Enterprise i:

It is expected that the export performance of an enterprise will increase, the better the enterprise complies with labor standards.

- Do you employ children below 16? (1=yes; 2=no)
- Is the worker in your firm eligible for maternity leave? (1=yes, 2=no)
- Compliance with "prohibition of child labor" (1=yes, 2=no)
- Compliance with freedom of association and protection of the right to organize (1=yes, 2=no)
- Compliance with minimum age (1=yes, 2=no)
- Do you think that increased market access will help abolishing child labor or will make you comply with labor standards? (1=yes; 0=otherwise)
- Is the worker in your firm eligible for paid vacation? (1=yes, 2=no)

3) Enforcement of Labor Standards from Outside the Enterprise:

We expect that the probability of an enterprise to export to the West with an export share of more than 50% will increase with the level of enforcement of labor standards. An exception is the variable related to the costs of standards for which we expect a negative correlation.

- In your opinion, what makes you comply with core labor standards? Is it an incentive in terms of better market access? (1=yes; 0=otherwise)
- Is it because of competition with other countries? (1=yes; 0=otherwise)
- In your opinion, what makes you comply with core labor standards? Is it because of inspections and monitoring by a domestic authority? (1=yes; 0=otherwise)
- Do you agree that these labor standards are very important to ensure fair competition? (1=yes; 0=otherwise)
- Are the requirements you follow in production based on mutual agreement between you and your major importing country? (1=yes; 0=otherwise)
- What costs do you think will or did occur as a percentage of total costs in the case of complying with these standards? (1=more than 5%; 0=otherwise)
- Do you think that imposing fines will stop child labor? (1=effective to very effective; 0=otherwise)
- Have there been any inspections in your company to control the use of child laborers by foreign or domestic authorities? (1=yes; 0=otherwise)
- Do you think that setting a minimum age for working will stop child labor? (1=effective to very effective; 0=otherwise)
- Are the requirements you follow in production imposed on you by local and / or foreign authorities? (1=yes; 0=otherwise)

Developing Country Responses to the Enhancement of Food Safety Standards

Spencer Henson and Steven Jaffee

1   Introduction

Food safety standards have become a more prominent issue for global trade in agricultural and food products (Jaffee and Henson 2004; Josling et al. 2004). Of particular concern is the potential impact of food safety standards on the ability of developing countries to both gain and maintain access to markets for high-value agricultural and food products, especially in industrialized countries. In part this reflects the growth of these standards, but also more widespread recognition of the degree and manner in which trade flows can be affected. Concerns are greatest in the case of low-income countries, given their typically weaker food safety and quality management capacities that might thwart efforts towards export-led agricultural diversification and rural development.

This paper explores the impact that food safety standards are having on the performance of developing countries with respect to agricultural and food product exports, drawing on a program of research work at the World Bank (see World Bank 2005). While recognizing that food safety and quality standards can act to impede exports, an attempt is made to 'rebalance' the policy debate in this area. The paper outlines how the proliferation and increased stringency of food safety standards are creating a new landscape that, in certain circumstances, can form a basis for the competitive repositioning and enhanced export performance of developing countries. In particular, the basis for this competitive repositioning is discussed and related, in turn, to the manner in which developing country governments and/or private sector suppliers respond to evolving standards.

2   Drivers of Food Safety Standards

The expansion of international trade in high-value agricultural and food products has served to highlight the extent to which national food safety standards diverge, as well as the differential capacities of both public authorities and private sector suppliers to comply. For many higher-value agricultural and food products, international competitiveness is no longer driven by price and quality grades (Jaffee and Henson 2004). Rather, safety concerns have come to the fore and the dominant modes of competition in many agricultural and food markets are based around quality rather than price (Busch and Bain 2004). There is greater scrutiny of the production or processing techniques employed along the associated supply chains (Buzby 2003; Unnevehr 2003) and a number of meta systems, for example hazard analysis and critical control point (HACCP), have increasingly become global food safety norms.

There are various reasons why food safety standards may differ between countries (Unnevehr 2003; Henson 2004). First, distinct tastes, diets, income levels and perceptions influence the tolerance of populations towards the potential risks associated with food. Second, differences in climate and the application of production and process technologies affect the incidence of different food safety hazards. Food safety standards, in turn, reflect the feasibility of implementing alternative mechanisms of control, which itself is influenced by legal and industry structures as well as available technical, scientific, administrative and financial resources. For example, some food safety risks are greater in developing countries due to weaknesses in physical infrastructure (for example efficacy of hygiene controls) and the higher incidence of certain infectious diseases. Further, climatic conditions may be more conducive to the spread of particular pests and diseases that pose risks to human health.

The intrinsic risks associated with the production, transformation and sale of agricultural and food products, combined with different standards and institutional capabilities, can pose major challenges for international trade. This is exacerbated by on-going and rapid changes in the landscape for food safety standards. Over the past decade, there has been increased public awareness and concern about food safety within industrialized countries in the wake of a series of highly publicized food scares or scandals (Henson and Caswell 1999). In some countries, these events have shaken the confidence of consumers in national systems of food safety regulation. In response, there have been significant institutional changes in food safety oversight and reform of associated regulations. For long-held concerns (for example, the potential environmental and health impacts of pesticides), there has been a tightening of standards in many countries. At the same time, new standards are being applied to address emerging and/or formerly unregulated hazards (for example, Bovine Spongiform Encephalopathy or heavy metals). Increased emphasis is being given to product or raw material traceability, plus increased resources have gone into border inspections of imported food products.

In parallel with the evolution of regulatory standards and oversight have been efforts by the private sector to address food safety risks and otherwise attend to the concerns and preferences of consumers and civil society (Henson and Reardon 2005). Much of the motivation behind this trend has been the mitigation of reputational and/or commercial risks. Further, for some products private food safety standards have become the basis of competitive processes of market differentiation. This has resulted in a rapidly expanding plethora of private standards and other forms of supply chain governance. While these efforts have been especially prominent among major food retailers, food manufacturers and food service chains in industrialized countries, such systems of private food safety governance are also being applied more widely in middle-income (and even some low-income) countries. This later phenomenon reflects, in part, the investments undertaken by multinational retail or food service chains and the broader development of the supermarket sector in low and middle-income countries (Reardon and Berdegue 2002).

3 Alternative Perspectives on the Trade Effects of Food Safety and Quality Standards

The proliferation and enhanced stringency of food safety standards has fomented considerable concern among low and middle-income countries and development agencies aiming to promote trade as a means to agricultural and rural development (see for example Henson et al. 2000; Unnevehr 2000; Wilson and Abiola 2003; Otsuki et al. 2001). Indeed, there is a widespread presumption that food safety standards are used as a protectionist tool, providing 'scientific' justifications for prohibiting imports of agricultural and food products, or discriminating against imports by applying higher and/or more rigorous regulatory enforced standards than on domestic suppliers. Such concerns have become heightened as traditional barriers to trade, for example tariffs, have been eroded through progressive rounds of multilateral trade negotiations. Even where standards are not intentionally used to discriminate against imports, there is concern that their growing complexity and the lack of harmonization between countries impedes the efforts of low and middle-income countries to gain access to potentially lucrative markets in industrialized countries.

There is also concern that many low and middle-income countries lack the administrative, technical and scientific capacities to comply with strict food safety standards, presenting potentially insurmountable barriers into the medium-term (Henson et al. 2000). Further, the associated one-off and recurring costs of compliance can undermine the longer-term competitive position of exporters and/or diminish the profitability of high-value agricultural and food exports. It is argued that the combined effects of these institutional weaknesses and costs of compliance costs contributes to the further marginalization of smaller and/or poorer countries and weaker economic players therein, including small-scale producers and micro and small enterprises (Wilson and Abiola 2003).

An alternative and less pessimistic view, however, emphasizes the potential opportunities provided by evolving food safety standards and the likelihood that certain developing countries can utilize such opportunities to their competitive advantage (Jaffee and Henson 2004; World Bank 2005). From this perspective, public and private standards are viewed, at least in part, as a necessary bridge between increasingly demanding consumer requirements and the participation of international suppliers. Many food safety standards provide a 'common language' through the supply chain, in turn reducing transaction costs, and promote consumer confidence in food product safety, without which the market for these products cannot be maintained and/or enhanced.

The costs of complying with food safety standards may also provide a powerful incentive for the modernization of export supply chains in low and middle-income countries. Compliance with stricter food safety standards can also stimulate capacity-building within the public sector and give greater clarity to the appropriate management functions of government. Further, through increased attention to the spread and adoption of 'good practices' in the supply of agricultural and food products, there may be spill-overs into domestic food safety systems, to the benefit of the local population and domestic producers. Thus, the associated costs of compliance are offset, at least in part, by an array of benefits, both foreseen and unforeseen, from the enhancement of food safety management capacity. Rather than degrading the competitiveness of low and middle-income countries, therefore, the enhancement of capacity to meet stricter food safety standards can potentially create new forms of competitive advantage. While there will inevitably be losers as well as gainers, this view suggests that the process of standards compliance can conceivably provide the basis for more sustainable and profitable agricultural and food exports in the long-term. In turn, it redirects the debate to the conditions under which developing countries are able to derive gains from evolving food safety standards.

This rather crude dichotomy between 'standards as barriers' and 'standards as catalysts' suggests a complex reality in which close attention is needed to the specifics of particular markets, products and countries to understand how food safety standards are providing challenges and opportunities for low and middle-income countries. Further, there is a need to understand the strategic options and patterns of performance of developing countries in meeting these challenges and their ability to exploit emerging opportunities. The following section provides a commentary on the varied concerns associated with standards and agricultural and food exports from low and middle-income countries, noting the availability of evidence that supports or opposes prevailing claims and the assumptions on which they are based. The result is a varied picture, partially supporting both of these opposing perspectives. In turn, this highlights the dangers of making over-ly generalized conclusions and the need to differentiate analyses and strategies in relation to food safety standards.

4   Food Safety Standards and Trade

During the Uruguay Round of multilateral trade negotiations, agricultural exporters voiced concerns that food safety, as well as animal and plant health measures (generally referred to as sanitary and phytosanitary or SPS measures) were sometimes used to restrict import competition to domestic producers and that such protectionist measures would likely increase as traditional trade barriers declined (Henson and Wilson 2005; Marceau and Trachtman 2002). The Agreement on the Application of Sanitary and Phytosanitary Measures (SPS Agreement) was negotiated in order to provide a set of multilateral rules that would both recognize the legitimate need for countries to adopt SPS measures and, at the same time, create a framework to reduce their potential trade-distorting effects.

The SPS Agreement built upon the Standards Code introduced in the 1947 General Agreement on Tariffs and Trade (GATT) (Marceau and Trachtman 2002). It permits measures that are 'necessary to protect human, animal or plant life and health', yet requires regulators to: (1) base measures on a scientific risk assessment; (2) recognize that different measures can achieve equivalent safety outcomes; and (3) allow imports from distinct regions in an exporting country when presented with evidence of the absence or low incidence of pests or diseases. In addition, the SPS Agreement encourages the adoption of international standards, making explicit reference to those of the Codex Alimentarius Commission (CAC) in the case of food safety. Importantly, the Agreement protects the right of a country to choose its own 'appropriate level of protection', yet guides members to minimize any associated negative trade effects (Henson 2001).

The SPS Agreement thus sets out broad ground rules for the legitimate application of food safety standards, many of which could affect international trade. Yet, the Agreement gives countries fairly broad latitude in setting and applying such measures. Scientific justification is called for wherever standards are deemed not to be based on established international standards. In practice, complications are inevitable given the wide range of areas for which no agreed international standards exist and given broad (and emerging) risks for which the state of scientific knowledge is incomplete (Roberts 2004). Hence, many of the controversies which have occurred surround the legitimacy and/or appropriateness of measures in the context of scientific uncertainty.

Important underlying objectives of the SPS Agreement are minimization of the protectionist and unjustified discriminatory use of standards and the promotion of greater transparency and harmonization. In both regards, experience has been mixed (Roberts 2004). The difficulties encountered are probably less due to specific shortcomings of the SPS Agreement itself, than the intrinsic complexities of the management of food safety protection and rapidly-evolving markets for agricultural and food products. Further, it is evident that WTO Members vary widely, both in their understanding of the Agreement and their ability to take advantage of the rights and responsibilities it defines.

The SPS Agreement has not eradicated the differential application of standards and it is, perhaps, unrealistic to expect it to do so. Indeed, differentiation in the application of food safety standards is a necessary part of any risk-based food safety control system. At the country, industry and enterprise levels, there is a need to prioritize the hazards to be monitored and associated control measures that are implemented, given resource limitations. Further, priorities are inevitably set, not only on the basis of scientific evidence, but also political factors, for example where consumers and other interest groups are showing most concern (Henson 2001). As resources are limited and the implementation of food safety standards is often costly, an effective risk management system will go beyond the prioritization of potential hazards to differentiate explicitly between alternative sources of supply based on distinct conditions of production, past experience and assessments/perceptions of risk management capabilities through the supply chain. Indeed, many countries operate systems of automatic detention for products imported from countries (or particular companies) with a history of non-compliance with food safety standards.

In circumstances where regulators have wide discretion and various forms of differentiation are required for cost-effective management of food safety, there remains scope for 'mischief'. Yet separating legitimate differenttiation from non-legitimate discrimination is problematic. It is even more difficult to attribute particular food safety standards to protectionist designs, considering that in most circumstances where protectionism is alleged, there are at least partially legitimate food safety concerns at play. The case of European Union (EU) standards for aflatoxins in nuts and cereals is a poignant example (see for example Otsuki et al. 2001a; 2001b). In other cases, trading partners have differing perspectives on the current state of scientific knowledge and/or the need to make allowance for uncertainty. Perhaps the most prominent case is the dispute between the EU and United States (US) over restrictions on exports of beef produced with the use of hormones (Paulwelyn 1999; Bureau et al. 1998).

Thus, there are remaining concerns over the degree to which there is systematic discrimination against imports in the application of food safety standards. One question is whether importers must comply with higher requirements than domestic suppliers. No systematic research has been undertaken on this subject, although a great deal of anecdotal evidence is presented by those that purport to have been adversely affected by food safety standards. Thus, 241 complaints were raised by WTO Members in the SPS Committee over the period 1995 to 2002 (Roberts 2004). On the basis of general impressions, it would appear that many countries, both industrialized and low and middle-income, do have a lower tolerance for food safety risks from imports than from domestic sources. For example, the US has long complained that a broad array of countries have a near zero tolerance for *Salmonella* in imported poultry products, yet this pathogen is widely present in the domestic supply chains of these countries (Jaffee and Henson 2004).

Currently, there is a paucity of systemic research that compares the modes and intensity with which food safety standards are enforced for domestic versus imported supplies. In discussions with high-value food exporters in low and middle-income countries, one frequently hears accusations that the controls they face are more rigorous than those imposed on domestic suppliers (Jaffee and Henson 2004). Frequently, however, this perception appears to emanate from the intensive oversight and monitoring provided by private entities, especially supermarkets and their buying agents, rather than from official systems of surveillance and product monitoring. Further, in many ways the methods of official control they can face are more 'visible' in their effects, in that compliance is assessed at the border and on this basis entry is possibly denied. Domestic suppliers, however, are typically regulated through inspection of their processing facilities with a focus on system-based controls and/or market surveillance.[1]

While it is not possible to denote generalized trends in relation to the justification for discrimination in the application of food safety standards, it is apparent that, at the very least, the transparency of official regulatory measures has improved in the period since the SPS Agreement entered into force. Around 85 percent of WTO members have established an 'Enquiry Point' as a conduit through which other WTO Members can obtain further information on proposed SPS measures. Between 1995 and 2002, WTO members submitted around 3,220 notifications of new SPS measures. These notifications provide advanced warning of new or modified measures and an opportunity for trading partners to raise questions /objections to the proposed measures, both bilaterally and through the SPS Committee. While it is evident that industrialized and developing countries may differ in their ability to respond to notifications, over time it is evident that an increasing proportion of WTO members, including developing countries, have taken advantage of this opportunity to raise their concerns (Roberts 2004).

While the notification process has increased the transparency of food safety standards, there remain considerable variations in standards between countries and widespread uncertainty over how certain countries are implementing/enforcing their standards. Roberts et al. (1999) note the paucity of international standards for many agricultural and food products. They indicate that, over the period 1995-1999, the vast majority of SPS measures notified to the WTO were ones for which no international standard existed. Jaffee (2003) notes that, despite efforts to harmonize Maximum Residue Levels (MRLs) for pesticides in fresh fruit and vege-tables imported into the EU, *de facto* there remain wide variations in operative standards due to different country approaches to surveillance and enforcement.

Variations in standards are also common in other sectors. Henson and Mitullah (2003) contrast the varied standards that low and middle-income countries must meet in order to gain and maintain access to US, EU and Japanese markets for fish and fishery products. While there are some

---

[1] There is also a paucity of systematic research comparing the intensity with which private buyers and distributors enforce their own food safety standards among domestic suppliers versus suppliers in other countries. Anecdotally, one would assume that they would have less opportunity to observe directly the food safety control systems employed by low and middle-income country suppliers and place particular emphasis on end-product testing and/or require that suppliers obtain (third-party) certification of their food safety management systems.

overlapping requirements, especially the increasing emphasis on application of HACCP, there remain significant differences in both regulatory and private requirements. Likewise, Mathews *et al* (2003) highlight the range of product and process standards required by countries to minimize the risk of *Salmonella* in poultry and poultry products. Dohlman (2003) and Otsuki *et al.* (2001) discuss the significant different-ces among countries, not only in the maximum permitted level for aflatoxins in cereals and nuts, but also the sampling methods used to assess conformity. This lack of harmonization in both standards and conformity assessment procedures can result in increased production and transaction costs for low and middle-income country suppliers, necessitating du-plicative testing and reducing their ability to achieve economies of scale in certain production or food safety management functions.

Two further trends are contributing to the increased complexity of the food safety standards environment. First, a growing number and proportion of food safety measures are risk-based process standards, relating to production, post-harvest and other procedures and/or the manner in which compliance is assessed. This reflects both the inefficiency and inefficacy of end-product testing, particularly in view of the levels of risk deemed acceptable and the emergence of 'new' food borne pathogens. Roberts (2004) notes that, over the past decade, the major international standards organizations have devoted more of their attention and resources towards the development of common approaches to risk identification, assessment, and management (i.e. meta-standards) than to international standards *per se*.

A second trend is the proliferation of private standards, encompassing both product and process specifications. Some of these are essentially food safety or food hygiene protocols, as with the British Retail Consortium (BRC) Technical Food Standard. Others combine a mixture of food safety, environmental and social dimensions, as exemplified by the most recent EUREPGAP Fruit and Vegetable Standard. These examples are all private protocols that have been codified and are available to the public (or at least to would-be suppliers). They represent attempts to harmonize varying food safety standards formerly applied by individual private companies. Yet, there still remains a plethora of private standards that are simply commu-nicated through individual supply chains and can vary widely in their specific requirements.

Continued variations in food safety standards alongside the progressive shift towards process-based measures have enhanced the importance of 'equivalence' of national standards and systems. A related issue is the mechanism through which equivalence is recognized, involving bilateral or multilateral agreements. Currently, there is no systemic recording of equivalence agreements although, at least anecdotally, those that have evolved appear to be between industrialized countries. However, even agreements between industrialized countries are limited and can take a great deal of time and effort to establish. For example, the Veterinary Equivalence Agreement between the US and EU took seven years to be established and arguably has had little tangible impact on differences in food safety requirements as they influence bilateral trade in livestock products. Certain low and middle-income countries, including those which have become highly successful agricultural and food exporters, have highlighted an array of difficulties in gaining recognition for the equivalency of their food safety and other controls to those of their major trading partners (WTO 2001). However, perhaps, one of more successful and wide-ranging example of 'equivalence' is the recognition by the EU that a broad range of developing and industrialized countries have established systems of hygiene control for fish and fishery products that offer a level of protection at least comparable to its own legislation (Henson and Mitullah 2004).

A parallel trend, reflecting the proliferation of private food safety standards, is the heightened importance of certification. Certification is the process by which buyers assess the compliance with defined standards and is typically undertaken by a third party agency that the buyer recognizes as 'competent'. In this context, a crucial issue for low and middle-income countries is the establishment of certification capacity and parallel institutions through which certification bodies are accredited. Exporters in countries that lack an accredited certification system may be forced to use the services of an accredited body in another country, most commonly an industrialized country, the cost of which can be considerable (El-Tawil 2002; WTO 2005)

While the process of notification under the SPS Agreement has contributed to increased transparency of official food safety standards, this has been accompanied by the proliferation of private standards that fall outside of the purview of the WTO. Thus, the overall picture for food safety requirements in international trade is becoming increasingly complex and dynamic as standards are promulgated in multiple spheres at industry, national, regional and international levels. Further, the complexity of this issue stems not only from the variability of standards on paper, it is magnified by differences in the ways, means and intensities by which the standards are monitored and enforced, which themselves are changing over time. Thus, for a developing country exporter, the operative 'rules of the game' are derived by a combination of factors including the prevailing standards themselves, enforcement capacities and predilections of official agencies, nature of private standards and oversight arrangements such as certification, and the prominence of particular concerns among consumers and civil society organizations at any point in time.

5   Food Safety Standards as a Strategic Issue

The complexity of the food safety standards environment highlighted above poses enormous challenges for low and middle-income countries in general, and stakeholders involved in export-oriented agricultural and food supply chains in particular. Embedded within these challenges, however, are a plethora of strategic decisions that policy-makers and private sector entities need to make in identifying the emerging set of requirements with which they must comply and any associated threats or opportunities. In so doing, they must trade-off the available options through which compliance can be achieved and manage the chosen processes of capacity-building and adjustment. The notion of 'strategic options' is quite novel in the context of food safety standards and trade, especially in the context of low and middle-income countries. The more typical assumption is that low and middle-income countries are 'standards takers', facing essentially all-or-nothing decisions regarding compliance with few, if any, alternative approaches to achieving their trade goals. The perspective presented here, however, focuses instead on the 'room for maneuver' available to low and middle-income countries in complying with food safety standards.

Figure 1 presents a simple conceptual framework that aims to characterize alternative strategic responses to food safety standards. This framework draws on the concepts of 'exit', 'loyalty' and 'voice' developed by Hirschman (1970). Hirschman's framework was originally used to examine economic and political behavior as responses to the decline of firms, organizations and states, but has since been extended to quite different contexts, for example microfinance for micro and small enterprises (Lepenies 2004). Depending upon the context, exit could involve leaving an organization, emigrating, or ceasing to buy a com-pany's products. Voice involves protest or otherwise lobbying for changes in rules and laws. For Hirschman, loyalty involves deepening one's participation in, and alignment with, an entity's goals and processes. A second 'pro-activity'-'reactivity' dimension relates to the time when efforts to comply commence, which is our own innovation.

The predominant dialogue on food safety standards, especially relating to low and middle-income countries, presents a single strategic option of complying with (official and private) food safety standards in focal markets, i.e. 'loyalty'. This can take a variety of forms, including the adoption of legal/regulatory reforms, changes in production technologies, shifts in the structure of supply chains, additional measures for conformity assessment, etc. This approach to compliance can be implemented at the time a standard comes into force, that is 'reactively', or ahead of time in view of expectations as to how standards are likely to evolve in the future, that is 'proactively'. Everything else being equal, a 'proactive' approach affords greater potential to manage compliance in a manner that brings about strategic gain. This relates to the existence of 'first mover' advantage, for example through earlier sunk costs or reputational effects, as well as the greater flexibility afforded by longer time periods over which compliance can be pursued. In a 'pro-active' mode, there is greater scope to test and apply alternative technologies and employ varied administrative and institutional arrangements.

	Reactive	Proactive
Exit	Wait for standards and give up	Anticipate standards, leave particular markets or market segments, and make other commercial shifts
Loyalty	Wait for standards and then adopt measures to comply	Anticipate standards and comply ahead of time
Voice	Complain when existing standards are applied or new measures are adopted	Participate in standard creation and/or negotiate before standards are applied

Fig. 1. Strategic response to food safety standards

In practice, however, there are other strategic options beyond 'loyalty'/ compliance. On the one hand, countries or individual private sector exporters can 'exit', choosing not to comply with the food safety standards being imposed in a particular market. This implies switching customers, in the case of a private standard, or exiting particular export markets altogether. The producer and/or exporter may choose to switch to different products for which the food safety (or agricultural health) standards are less problematic or costly, for example certain processed rather than fresh products or meat rather than live animals. Such a strategy might be employed where compliance will yield a fundamental loss of competitiveness and/or negative economic and social impacts, where resources might be better spent elsewhere, and/ or where profitable alternative markets exist that have less demanding standards, for example the higher quality segments of domestic markets or in other developing countries. Thus, 'exit' should not be construed as a loser's strategy; it can take the form of a carefully considered re-direction of commercial strategy.

In parallel with strategies of 'loyalty' or 'exit', low and middle-income country governments and/or exporters can adopt a strategy of 'voice', seeking to influence the prevailing rules or responding to new standards by negotiating or complaining. For example, WTO members may raise their complaints through a cross-notification in the SPS Committee or engage in bilateral negotiations with their trading partners regarding the specific actions required to achieve compliance. Individual exporters may question the food safety standards being imposed by their customers and attempt to come to some compromise that reflects their own circumstances alongside customer's demands. Across both 'exit' and 'voice', being 'proactive' is considered more strategically advantageous than being 'reactive'. Typically in any one industry, a combination of all three types of strategies is likely to be observed, yet in differing proportions and perhaps involving different stakeholders.

Besides the two dimensions in Figure 1, there are further ways to characterize the responses of low and middle-income countries to new food safety standards in export markets. One distinction is between 'defensive' and 'offensive' approaches. 'Defensive' strategies are aimed at maintaining the status quo and minimizing related impacts. The aim is normally to limit the actions (and often also the investments) needed to achieve compliance. This is often pursued under conditions of resource limitations and risk adversity. 'Offensive' strategies involve attempts to utilize standards as a means to gain competitive advantage, even where this may require additional investments beyond the minimum required to achieve compliance.

A final dimension relates to the locus of strategic response. Measures can be taken within the public or private sectors, involving either individual entities (for example single exporters or producers) or various forms of collective action. Where both the public and private sector are adopting measures, the leadership or driving force behind this process could come from either side. Traditionally, relatively clear distinctions have been made between aspects of food safety management that are the domain of the public and private sectors. Increasingly, however, these demarcation lines are being challenged. For example, the potential role of self-regulation through industry-level 'codes of practice' and commercial laboratories for product certification is being acknowledged. Further, there is recognition of the potential efficiencies associated with collective and collaborative actions. These can include inter-ministerial task forces seeking to avoid duplication of efforts where multiple tiers of government are involved and/or trade and industry associations that build on the compliance investments made by individual enterprises. Collective action can also take place across the public and private sectors, for example through joint task-forces. More broadly, it is recognized that both the public and private sectors have a role to play in responding to new food safety standards, and that national standards capacity should be viewed from this holistic perspective.

In the context of this framework, the most positive and potentially advantageous strategy combines 'voice', 'proactivity' and 'offensive' orientations. Everything else being equal, this approach is most likely to turn the challenges associated with new food safety standards into a competitive opportunity and to yield positive social and economic benefits. Conversely, the most negative approach is a combination of 'exit', 're-activity' and 'defense'. Indeed, there may be considerable costs associated with such an approach related to sunk investments, and the social and economic consequences for supply chains that are export-oriented. In turn, the strategic opportunities available to countries and/or exporters within countries will reflect prevailing capacities, specifically related to SPS management but also more generally, the nature and *modus operandi* of supply chains, nature of specific SPS standards, etc. In this context, the focus of capacity-building should be on the enhancement of strategic options.

6 Strategic Approaches to Food Safety Standards in Developing Countries

In examining the strategic response to evolving food safety standards by low and middle-income countries, a distinction is made between the ways in which countries have reacted to new standards at the international level, for example through the WTO, and the specific compliance efforts of both the private and public sectors. While far from exhaustive, these provide some salience to the strategic perspective being presented here. Each is discussed in turn below, in the case of specific compliance responses through the examples of fish, horticultural and spice exports from India and Kenya.

6.1 International 'Voice'

An indicator of the degree to which developing countries are able to exhibit 'voice' when new food safety standards are proposed by trading partners is provided by the number and nature of complaints and counter-notifications made through the SPS Committee. Admittedly, this is a rather 'reactive' mode of 'voice', as discussed above, but our analysis is constrained by the non-availability of data on other responses, for example bilateral complaints and negotiations. Table 1 provides a summary of the pattern of counter-notifications according to regulatory goal (covering not only food safety but also plant and animal health) and the country group raising the issue or being the subject of a complaint (Jaffee and Henson 2004; World Bank 2005). These data suggest that low and middle-income countries have used the formal review and complaint processes of the SPS Committee quite actively since its inception in 1995 to register their concerns with respect to a significant number of notified measures, both by industrialized and other low and middle-income countries. A more detailed look at the individual complaints, however, yields a more complex picture, as described below.

Complaints by developing countries are dominated by a small number of middle-income countries, in particular Argentina, Brazil, Chile and Thailand. Each of these countries has issued or supported multiple complaints. These four countries have been involved, in one way or another, in the vast majority of complaints by low and middle-income countries. Very few other low and middle-income countries have been involved in multiple cases. This pattern of participation reflects the prominence of certain countries in the trade of a few product categories, especially beef and horticultural products, rather than the overall structure of low and middle-country agricultural and food trade. Low-income countries are weakly represented in the pool of counter-notifications, issuing or supporting complaints in only five cases. This could partly be a reflection of the structure of their exports, which are concentrated in commodities for which SPS measures are of lesser importance, or their limited capacity and/or confidence to participate in the SPS Committee. The lack of formal complaints by low-income countries is, however, no reflection of their ability to resolve effectively their concerns bilaterally. Thus, these data alone provide us with very little information regarding the extent to which SPS measures are inhibiting the exports of low-income countries.

Food safety-related complaints account for half of all counter-notifications of SPS measures. These are a mixture of quite specific concerns with no large clustering around any particular theme. The rationale behind counter-notifications related to food safety standards is predominantly the purported 'lack of scientific evidence'. Among low and middle-income countries, the EU has been the subject of the largest number of complaints related to food safety. For example, there were more than three times as many complaints against the EU than against the US over the period 1995 to 2003. Several reasons might account for this. First, the process of harmonization of SPS measures within the EU has often resulted in the adoption of the most stringent standards previously applied by individual Member States. Second, the EU has more frequently and most visibly embraced the 'precautionary principle' when adopting food safety standards, sometimes giving rise to controversies over the scientific basis for its actions. Third, due to the complex administrative structure of the EU, some countries reportedly find it difficult to resolve concerns through bilateral discussions and therefore resort more readily to the venue of the SPS Committee to take up concerns with the European Commission.

The growing number of recorded complaints or counter-notifications by developing countries, however, provides only a crude indicator of the extent to which they are able and willing to exhibit 'voice'. These complaints probably represent the 'tip of the iceberg' with a greater proportion of concerns and disputes being raised bilaterally. At the same time, however, it could also indicate that low and middle-income countries in general lack the capacity to complain or negotiate when new food safety standards, as well as SPS measures more broadly, are applied. Further, the apparatus of formal complaints through the WTO relates only to mandatory standards set by public agencies. As described above, a growing array of food safety standards are being set privately, either through consensus within particular industries or by the 'gate keepers' of the dominant supply chains. While many such standards are ostensibly voluntary, they are becoming the *de facto* standards with which compliance is required to gain or maintain access to particular buyers or market segments. 'Voice' relating to these standards will occur through the private bilateral negotiations between supplier and customer. These private negotiations cannot be empirically aggregated.

Data are available on developing country participation in international standards-setting organizations in the area of food safety, notably the Codex Alimentarius Commission. These data provide some evidence of the degree to which low and middle-income countries are able to exhibit 'voice' at the international level through participation in international standards development. Around 80 percent of developing countries are members of Codex Alimentarius (Henson et al. 2001; Henson 2002). However, their participation in the Codex Alimentarius Commission itself, which ratifies all new standards, remains limited. Thus, in 2004 only 39 percent of low and 47 percent of middle-income country Members of

Codex attended the Commission meeting. Indeed, regular participation in Codex Alimentarius is typically limited to a group of larger and/or middle income countries including India, China, South Africa, Brazil, Argentina, Mexico, Malaysia, Thailand and Chile. While some other low and middle-income countries, for example Kenya and Egypt, have made efforts to enhance their participation, most countries attend meetings irregularly at best. Further, standards development itself takes place in a series of General Purpose and Commodity Committees that generally meet on an annual basis. Low and middle-income country participation in these meetings is typically very low, suggesting that, even where they do participate in Codex, it is very much in a 'reactive' mode.

Table 1. Counter-notifications relating to new measures in the SPS committee, 1995-2003

Complaints Against Measures of	Regulatory Goal of Contested Measure				
	Plant Health	Animal Health	Human Health	Other*	Total
	Number of Complaints by Developed Countries				
Industrialized Countries	18	11	49	3	81
Low/middle-income Countries	19	15	41	4	79
Multiple Countries	-	2	1	-	3
Sub-total	37	28	91	7	163
	Number of Complaints by Developing Countries				
Industrialized Countries	14	14	38	2	68
Low/middle-income Countries	8	19	7	2	36
Multiple Countries	1	2	-	-	3
Sub-total	23	35	45	4	107
Total	60	63	136	11	270

* Includes complaints about horizontal regulations (such as those regulating products of modern biotechnology) that reference human, animal, and plant health as objectives.
Source: Jaffee and Henson (2004) updating Roberts (2004)

In conclusions, it is evident that many low and middle-income countries face considerable constraints that limit their participation in both the SPS Committee and Codex Alimentarius which, in turn, mutes their international 'voice'. In many cases, the necessary resources are not available to attend multiple meetings each year, most of which are in industrialized countries. In the case of the WTO, a number of smaller low and middle-income countries do not even have permanent missions in Geneva. Further, even where attendance at meetings is possible, many countries lack the technical know-how, background scientific data and/or experience to utilize these fora to address their interests and concerns related to food safety standards.

6.2 Some Case Studies

More concrete and in-depth evidence of strategic approaches adopted by developing countries in complying with food safety standards for agricultural and food products in international trade can be provided by, and in fact requires, in-depth case studies (World Bank 2005). Here the cases of fish, horticultural, and spice exports from India and Kenya to the EU are presented as illustrative examples (for more in-depth analysis see Henson and Mitullah 2004; Henson et al. 2005; Jaffee 2003; Jaffee 2005).

Fish and Fishery Products

Over the last decade, developing country exports of fish and fishery products have increased at an average rate of six percent per annum (Delgado et al. 2003). However, one of the major challenges facing low and middle-income countries in seeking to maintain and expand their share of global markets is progressively more strict food safety requirements, particularly in major industrialized countries. Previous studies suggest that exporters in a number of countries have experienced considerable problems complying with these requirements (See for example Henson et al. 2000; Rahman 2001; Musonda and Mbowe 2001; UNEP 2001a; 2001b; Zaramba 2002). While the associated costs of compliance can be significant, however, the returns in terms of continued and/or expanded access to high-value markets often more than compensates (Cato and Subasinge 2004; Ponte 2005).

The EU lays down harmonized requirements governing hygiene throughout the supply chain for fish and fishery products. Processing plants are inspected and approved on an individual basis by a specified 'Competent Authority' in the country of origin, whether an EU Member State or a Third Country, to ensure that they comply. The European

Commission undertakes checks to ensure that the Competent Authority undertakes this task in a satisfactory manner. Imports from Third Countries are required to have controls that are at least equivalent to those of the EU[2]. Countries for which local requirements have been recognized as equivalent are subject to reduced physical inspection at the EU border. Countries that have not yet met these requirements, but which have provided assurances that their control are at least equivalent to those of the EU, are currently permitted to export, subject to higher rates of border inspection. Initially the deadline for all countries to be fully-harmonized with the EU's hygiene standards was December 31, 1996. However, this has been extended on numerous occasions and the current deadline is December 31, 2005.

While India and Kenya differ in terms of the specific products exported - India mainly exports shrimp, squid and cuttlefish, while Kenya's exports are dominated by Nile perch – they share common experiences with enhanced food safety standards. Both provide examples of longer term efforts to comply with the EU's hygiene standards for fish and fishery products, overlaid with the necessity to overcome restrictions on trade relating to immediate food safety concerns. In the case of Kenya, restrictions related to general hygiene standards in processing establishments alongside specific concerns relating to microbiological safety and pesticide residues were applied on-and-off over the period 1997 to 2000. India was subject to similar restrictions related to hygiene standards in fish processing during 1997. In both cases the restrictions served to significantly restrict access to EU markets.

In both India and Kenya the dominant strategic approaches to emerging food safety standards have been 'reactive', 'loyal' and 'defensive', both by government and the private sector. Thus, hygiene and/or antibiotic controls have been largely up-graded in response to regulatory change in the EU and the demands of major customers. Further, in the cases of Kenya, little action was taken until inspection visits by the European Commission, which led to restrictions on imports to the EU. In India's case, the government had undertaken some initial reforms to its regulatory framework, although these were insufficient to comply with the EU's requirements. In both cases the substantive drive to up-grade hygiene controls occurred suddenly.

---

[2] The European Commission has presented its controls on hygiene for imports of fish and fishery products as a practical example of the application of equivalence (WTO 2002). Thus, rather than laying down specific requirements, the Commission focuses on the conditions under which products will be equivalent to those produced in the European Union.

Across both India and Kenya there were examples of exporters that adopted 'proactive' and 'offensive' strategies; these firms had seen the drive towards higher food safety standards and had made substantive efforts to up-grade their controls in a bid to meet these standards ahead of their competitors. While in most cases these represented a relatively small part of the total industry, they clearly stuck out as industry leaders. At the same time, however, there were exporters that had exited the industry in response to the imposition of stricter food safety controls; some withdrew from the business altogether, while other processors re-focused towards markets with lower food safety standards. Standards-related pressures were not the sole factors in this exit. Other on-going issues, including resource management and broader competitive and capacity pressures, served to exacerbate the impact of needed investments in order to comply with the new food safety standards. All of these firms had exited in a 'reactive' and 'defensive' manner.

In both India and Kenya there were some attempts to implement 'voice', although this has been in a 'responsive' and 'defensive' mode in response to restrictions already imposed or threatened by the EU. Both the government and industry were involved in such efforts, which clearly were designed to 'put out fires'. While on-going negotiations may have taken place between individual exporters and their customers, none of the exporters interviewed as part of the case studies alluded to these, suggesting that they were not a major element of strategic responses to evolving standards.

Horticultural Products

Over the past 30 years, developing countries have experienced rapid growth in their exports of fresh produce, mainly consisting of fruit and vegetables. This trade has spread from an initial base of traditional tropical fruits to include a broad array of products, stimulated by growing consumer interest in health and demand for fresh produce variety, freshness and year-round availability. At the same time, this trade has been facilitated by advances in production, post-harvest and cold chain logistical technologies and by increased levels of international investment. On every continent there have been notable 'success stories' in this field alongside a range of other countries which have struggled to maintain or improve their positions in international markets. This reflects the highly competitive and rapidly-changing nature of the industry, with multiple factors impacting on competitiveness (Diop and Jaffee 2004).

The regulatory and private governance systems for international fresh produce markets are becoming increasingly complex. This changing regulatory environment appears to be raising the bar for new entrants while throwing new challenges in the path of existing developing country suppliers. Many analysts and practitioners are expressing concern about the inability of small and/or low-income countries to meet rising public and private standards, and thus their capacity to remain competitive in international fresh produce markets. (Dolan and Humphrey 2000; Chan and King 2000; Buurma *et al.* 2001). Certain high-profile food scares and highly publicized instances of violative levels of pesticide residues have created an impression of extreme vulnerability on the part of developing country suppliers. Yet, experiences are mixed; Kenya's recent experience is one of absolute and relative success, reflecting either 'proactive' or 'reactive' approaches towards compliance/'loyalty' that have been aimed at exploiting real or perceived strategic gains.

Kenya's fresh produce trade dates to the mid-to-late 1950s, when small quantities of temperate vegetables and tropical fruits were supplied in the European winter 'off-season' to up-market department stores in London. This off-season trade continued and was later joined by year-round-supplies of high-quality green beans and a broad array of vegetables that comprised part of the traditional diets of the UK immigrant population from South Asia. Most of these products were air-freighted in small boxes for sale through wholesale markets or to distributor/caterers.

For many years, the industry functioned with very simple supply chains, involving little investment in infrastructure, product development or management systems. Around 12 medium-sized firms alongside large numbers of small, part-time operators handled the exports, frequently trading with relatives or similarly small-scale companies in Europe. Fresh produce was purchased from large numbers of small and larger growers. Produce was generally collected from farms or along roadsides, from where it was brought to a basic central warehouse, sifted and re-graded if necessary, cooled a little and trucked to the airport for shipment in the evening. Some limited inspection of produce was undertaken by Ministry of Agriculture officials at the airport. With relatively few exceptions, this was more or less the 'model' within the industry from the 1960s through to the mid-to-late 1980s. The Kenyan fresh produce industry remained competitive in some markets and for some products, but not for others. While experiencing some growth in the 1970s, the fresh produce exports from Kenya more or less stagnated in the 1980s.

Since the early 1990s, however, the Kenyan fresh produce industry has been reshaped and transformed, both 'proactively' and 'reactively', in response to and in anticipation of commercial, regulatory and private governance changes within its core external markets. Commercial pressures came in the form of saturated markets for certain products and increased competition from various suppliers which had improved their supply capabilities and had less expensive sea or air-freight costs than did Kenya. Commercial changes within Europe also required a shift in the dominant approach. In many countries, large supermarket chains were in ascendancy while wholesale markets were declining in importance or taking on more specialized roles. Consolidation was also occurring among importers, packers and distributors. Hence, the growing segments of the fresh produce market were being managed by fewer players. On the regulatory front, there was a steady wave of activity geared toward strengthening and harmonizing EU and Member State regulations and monitoring systems for food safety, quality conformity and plant health. Interspersed in this wave of regulatory activity were progressively-refined private sector standards (or 'codes of practice') governing food safety, among other things, plant health.

Several of the leading Kenyan exporters caught an early glimpse of this 'new world' fresh produce market and began to re-orient their operations in an 'offensive' manner. With the encouragement of several UK super-markets, they began to experiment with new crops. New consumer packaging was introduced and different combinations of vegetables were included. An increasing proportion of product was directed to selected supermarket chains. The latter began to send 'audit' teams to Kenya to check hygiene and other conditions on farms and in pack-houses. Im-provements and investments were recommended, and in some cases required. With renewed confidence in the future of the industry, several exporters made considerable investments in new or up-graded pack-houses and related food safety management systems for the packing of ready-to-eat, semi-prepared products. Systems for crop procurement have also been transformed with many of the leading companies investing in their own farms and/or inducing major changes in the production practices of out-growers. There has been an array of joint public/private sector initiatives to train growers in all aspects of 'good agricultural practice'. Through both 'reactive' and 'proactive' offensive strategies of 'loyalty'/compliance, Kenya thus moved beyond being a commodity supplier, with mixed salads, stir-fry mixes, vegetable kebabs and other value-added products now accounting for more than 40 percent of what has been a burgeoning trade over the past decade. Between 1991 and 2003, Kenya's fresh vegetable exports increased from $23 million to $140 million.[3]

---

[3] Not all of the industry has transformed itself. There remain around 25 smaller exporters who lack the financial resources to invest in modern pack-houses and continue to supply 'loose' produce to commission agents and others in European wholesale markets and the Middle East.

Rising private sector and public standards have posed challenges to the Kenyan fresh produce industry, yet at the same time they have also thrown a 'life line' to the industry. Due to its location and relatively high air-freight costs, the Kenyan fresh produce sector cannot compete with many other players on a unit-cost basis. Margins have been squeezed in the market for mainstream and 'commodity' vegetables. With rising labor costs in Europe, the Kenyan industry has repositioned towards higher level; of preparation, including sliced vegetables and salads, which involve labor-intensive functions. To date, this market segment has grown fastest in the UK, although there is increased buyer interest and consumer demand in the rest of Europe. This suggests that well-organized industries in low-income countries can indeed use stricter standards as a catalyst for change, and profit in the process.

Spices

Historically, international trade in spices was governed by a system of quality grades and cleanliness parameters. Since the early 1990s, however, health and hygiene specifications have gradually been incorporated into commercial spice supply chains and, to a lesser extent, into official regulatory systems. The vast majority of these product and process standards were not designed specifically for spices, but derive from general food standards related to microbiological contamination, pesticides, food additives, and food labeling. The changing commercial and regulatory requirements are well illustrated by the case of dried chillies and the challenges posed to India's continued supply of this product to the EU market.

Chillies are one of the few spices produced in India for which agro-chemicals are commonly used. Chillies are vulnerable to a variety of pests and diseases and are commonly grown in rotation with other commercial crops. While there have been periodic concerns or campaigns to address the risks that agro-chemicals pose to farmers and agricultural workers in India, there has not, until recently, been much mention of pesticide residue concerns in spices. This began to change in the early 1990s in the context of the broader program within the EU to harmonize the permissible MRLs in food products. Initially, questions were raised on spices by regulators and/or buyers in Germany. In 1994/1995, several consignments of Indian dried chillies were rejected by Spanish authorities because the detected pesticide residues exceeded the permissible MRLs for fresh/green chillies. In the late 1990s, additional consignments of Indian chillies and other spices were rejected in Europe and elsewhere, frequently because no established tolerance level existed for particular pesticides and spices.[4]

India's response to this challenge has combined elements of 'voice', 'loyalty' and 'exit', mostly in a 'reactive' mode. For example, the industry there has sought to influence the prevailing 'rules of the game'. Working in conjunction with various other country spice trade associations, the India Spices Board and the All India Spice Exporters Forum established an International Organization of Spice Trade Associations (IOSTA), which obtained observer status at the CODEX Committee on Pesticide Residues. The IOSTA has actively sought to gain recognition of new MRLs based on monitoring (rather than the more expensive field trial) data and acceptance of multiplication factors for MRLs for spices which are the dried form of vegetables for which established MRLs exist (i.e. for pepper, garlic, onion).

In parallel to this exercise of 'voice', the Indian spice industry has made various adjustments to comply with EU Member State requirements, even though such countries continue to account for only a small proportion of India's total exports of chillies. Among the measures taken have included a program of supervised field trials to establish a wider range of national MRLs, extension programs in major production areas to encourage a-doption of integrated pest management practices and/or promote organic production of chillies and public and private sector investments in la-boratory equipment to test chillies for a broader range of agrochemical residues. Contract farming arrangements have also evolved in which ex-porters provide seeds, detailed pest management guidelines, supervisory help (and policing) and premium prices for pesticide residue-free supplies. Exporters have also undertaken increased screening of intermediary ven-dors, giving preference to those which maintain proper purchasing records and provide oversight on farmer production practices.

'Exit' has also been a strategy pursued by certain Indian spice exporters. These firms have withdrawn from selected European markets and have re-directed their sales of chillies to other markets, especially in developing countries. While Indian exports of chillies to Europe have been stagnant in recent years, exports to developing countries have experienced very sharp increases. While little attention is given to pesticide residue matters by buyers or regulators in these other developing countries, some of the mea-sures taken by the Indian industry have improved its overall competi-

---

[4]   There exist only a handful of CODEX standards for MRLs related to agro-chemical use on spices. Individual countries have set MRLs themselves, generally for particular spices that are grown domestically in small quantities. Most of the spice and pesticide MRLs which do exist vary between countries.

tiveness in those markets. There are also small but growing consumer segments within the large Indian domestic market for spices that are demanding more 'safe' and 'sustainable' production practices.

7   Conclusions

This paper has put forward and examined the concept of 'standards as catalysts' in the context of food safety standards in international trade and the 'room for maneuver' that low and middle-income countries may possess in the face of an ever-changing and increasingly complex standards environment. This contrasts with the 'standards as barriers' perspective that has dominated the literature on food safety standards and agricultural and food trade. In so doing, however, the aim has not been to deny that food safety standards do not sometimes impede agricultural and food exports from low and middle-income countries. Rather, the dominant theme is the need for a strategic orientation when considering the trade effects of food safety standards.

This paper has presented evidence that is both limited in its scale and scope. However, it illustrates the range of strategic approaches employed by low and middle-income countries, both at the level of nation states in challenging regulatory standards and/or participating in international standards-setting. Further, the paper highlights the specific actions taken at the country and/or exporter levels when faced with enhanced food safety standards. These illustrate the ways in which strategic responses vary a-cross countries and between exporters therein, reflecting prevailing capacities and perspectives on emerging standards. Overall, these responses are typified by strategies that are 'reactive' and 'defensive'. At the same time, however, there are exporters that are 'proactive', complying ahead of their competitors and often deriving competitive advantage as a result. Across these various scenarios there is evidence of 'voice', although it is less evident that this has a major 'pay-off', while efforts in this regard are severely curtailed by capacity constraints.

An important implication of the strategic perspective presented here is the need for capacity-building efforts related to food safety to be recast away from the conventional focus on problem-solving and coping strategies, often centered on the development of technical infrastructure. Instead, capacity-building should be geared towards maximizing the strategic options available to both government and the private sector in low and middle-income countries when faced with new or more stringent food safety standards and enhancing their ability to employ strategies that generate gains in terms of export competitiveness.

References

Bureau J, Marette S, Schiavina A (1998) Non-tariff trade barriers and consumers' information: the case of the EU-US trade dispute over beef. European Review of Agricultural Economics 25: 437-462

Buzby J (ed) (2003) International trade and food safety: economic theory and case studies. United States Department of Agriculture, Agricultural Economics Report No 828, Washington DC

Delgado CL, Wada N, Rosegrant MW, Meijer S, Ahmed M (2003) Fish to 2020: supply and demand in changing global markets. International Food Policy Research Institute, Washington DC

Dohlman E (2003) Mycotoxin hazards and regulations: impacts on food and animal feed crop trade. In: Buzby J (ed) International trade and food safety: economic theory and case studies. United States Department of Agriculture, Agricultural Economic Report No 828, Washington DC

El-Tawil A (2002) An in-depth study of the problems by the standardizers and other stakeholders from developing countries - ISO/ WTO regional workshops: part 1, International Organization for Standardisation, Geneva

Henson SJ (2001) Appropriate level of protection: a European perspective. In: Anderson K, McRae C, Wilson D (eds) The economics of quarantine and the SPS agreement. University of Adelaide, Centre for International Trade Studies, Adelaide

Henson SJ (2004) National laws, regulations, and institutional capabilities for standards development. Prepared for World Bank training seminar on Standards and Trade, January 27-28, 2004 Washington, DC

Henson SJ, Loader RJ, Swinbank A, Bredahl M, Lux N (2000) Impact of sanitary and phytosanitary measures on developing countries. University of Reading, Department of Agricultural and Food Economics, Reading

Henson SJ, Mitullah W (2004) Kenyan exports of Nile perch: impact of food safety standards on an export-oriented supply chain. Policy Research Working Paper 3349, World Bank, Washington DC

Henson SJ, Preibisch KL, Masakure O (2001) Enhancing developing country participation in international standards-setting organisations. Department for International Development, London

Henson SJ, Saqib M, Rajasenan D (2005) Impact of sanitary and phytosanitary measures on exports of fishery products from India: the case of Kerala. World Bank, Washington DC

Hirschman AO (1970) Exit, voice, and loyalty: responses to decline in firms, organizations, and states. Harvard University Press, Cambridge

Jaffee S (2003) From challenge to opportunity: transforming Kenya's fresh vegetable trade in the context of emerging food safety and other standards in

Europe. World Bank, Agriculture and Rural Development Discussion Paper No 2, Washington, DC

Jaffee S, Henson SJ (2004) Food exports from developing countries: the challenges posed by standards In: Aksoy MA, Begin JC (eds) Global agricultural trade and developing countries. Oxford University Press, Oxford

Jaffee S, Henson SJ (2004) Standards and agro-food exports from developing countries: rebalancing the debate. Policy Research Working Paper 3348 World Bank, Washington DC

Josling T, Roberts D, Orden D (2004) Food regulation and trade: toward a safe and open global system. Institute for International Economics, Washington DC

Mathews K, Bernstein J, Buzby J (2003) International trade of meat/poultry products and food safety issues. In: Buzby J (ed) International trade and food safety: economic theory and case studies. United States Department of Agriculture, Agricultural Economic Report No 828 Washington, DC

Musonda FM, Mbowe W (2001) The impact of implementing SPS and TBT agreements: the case of fish exports to European Union by Tanzania. Consumer Unity & Trust Society, Jaipur

Otsuki T, Wilson J, Sewadeh M (2001) Saving two in a billion: quantifying the trade effect of European food safety standards on African exports. Food Policy 26:495-514

Pauwelyn J (1999) The WTO agreement on sanitary and phytosanitary (SPS) measures as applied in the first three SPS disputes. Journal of International Economic Law 94:641-649

Rahman M (2001) EU ban on shrimp imports from Bangladesh: a case study on market access problems faced by the LDCs. Consumer Unity & Trust Society, Jaipur

Reardon T, Berdegue JA (2002) The rapid rise of supermarkets in Latin America: challenges and opportunities for development. Development Policy Review 20:371-388

Roberts D (2004) The multilateral governance framework for sanitary and phytosanitary regulations: challenges and prospects. Prepared for World Bank training seminar on Standards and Trade, January 27-28, 2004 Washington, DC

Roberts D, Josling T, Orden D (1999) A framework for analyzing technical trade barriers in agricultural markets. United States Department of Agriculture, Economic Research Services, Washington DC

UNEP (2001) Country case studies on trade and the environment: a case study of Bangladesh's shrimp farming industry. United Nations Environment Programme, Geneva

UNEP (2001) Country case studies on trade and the environment: a case study on Uganda's fisheries sector. United Nations Environment Programme, Geneva

Unnevehr L (2000) Food safety issues and fresh food product exports from LDCs. Agricultural Economics 23:231-240

Unnevehr L (2003) Food safety in food security and food trade: overview. In: Unnevehr L (ed) Food safety in food security and food trade. International Food Policy Research Institute, Washington DC

Wilson J, Abiola V (2003) Standards and global trade: a voice for Africa. World Bank Trade and Development Series, Washington DC

World Bank (2005) Challenges and opportunities associated with international agro-food standards. World Bank, Washington DC

Zaramba S (2002) Uganda country report on the integration of multiple sources of technical assistance to capacity building on improving the quality of fish for export. Food and Agriculture Organization of the United Nations, Rome

Scope and Limitations for National Food Safety and Labeling Regimes in the WTO-Frame

Bettina Rudloff

1 Introduction

This chapter will discuss whether there is scope for a sovereign design of domestic food policies within the WTO-frame. In a first step, the existing scope provided will be described and secondly, the actual use of this scope based on findings of closed disputes will be analysed. This survey on real cases will be split into the period before the SPS-Agreement was adopted in 1994 and into the period afterwards. The main emphasis lies on standards but analogies for labels can be made as they are referring to underlying standards.

All food safety measures can be analysed in the framework of non-tariff barriers (NTBs). According to Gandolfo's definition of NTBs such measures are different from tariffs and cause negative trade effects (Gandolfo 1998). The latter attribute defines the major rationale for WTO rules on NTBs. Food safety measures may become an NTB as far as they are not just domestically implemented but applied to imports as a precondition for market access (Bagwell and Staiger 2002, p. 126). NTBs are addressed by different WTO provisions:

- The General Agreement on Tariffs and Trade (GATT) defines some general rules in Article I, III, IV and XX,
- Certain Agreements specify these GATT rules for selected issues like the Agreement on Technical Barriers to Trade (TBT-Agreement), that is addressing all technical regulations for products, and the Agreement on the Application of Sanitary and Phytosanitary Measures (SPS-Agreement) on food safety issues.

2 Time Prior the Adoption of the SPS-Agreement

Prior to the adoption of the TBT-Agreement and the SPS-Agreement in 1994 all emerging food cases were ruled on basis of GATT principles, namely the most-favoured nation principle (GATT Article I) and the national treatment rule (GATT-Article III).

Both principles command that "like products" must not be treated differrently, neither when comparing imports originating in different countries nor when comparing imports with domestic products. Hereby, no barrier neither a tariff nor a NTB on "like" products would be allowed:

- The most-favoured nation principle prohibits discrimination between imports of "like products" originating in different countries. Accordingly, this implies that discrimination of "unlike products" is possible.
- According to national treatment, domestic fees or rules can be applied to "like" imports only as far as they do not lead to worse treatment compared to domestic products. This implies for "unlike products" a potentially different treatment.

As conclusion, import barriers may be justifiable by Articles I and III only as far as unlike products are concerned. The interpretation of likeness of products affected by an accused trade barrier was centric in several disputes.

Table 1. The "like concept" in agricultural and food related disputes (1950 – 1994)

All closed disputes referring to agriculture and food	45	
Cases on interpreting the "like concept"	12	
Findings in favour of "unlike products"	2	• Case on support of domestic feed proteins: different feed proteins accepted as unlike (US against EC, BISD 25S/49)
		• Case on tariffs on wood types: different wood types accepted as unlike (Canada against Japan, BISD 36S/167)

Source: Own calculation on basis of the published cases at
http://www.wto.org/english/tratop_e/dispu_e/dispu_subjects_index_e.htm
(Dec 2004).

Out of 45 cases related to agriculture and food about one third was relating to the interpretation of "likeness" of the affected products to reject or justify a NTB at stake (Table 1). Just in two cases, the findings were in favour of "unlike products" and thereby, the respective import barriers could have been accepted. This acceptance was based on detectable physical attributes like different wood or protein types. Only in these two cases, the challenged barriers were evaluated as being in line with Article I and III, whereas in all other cases, the barrier at stake needed to be abolished according to the dispute findings.

The following Figure 1 shows a systematisation of the underlying attributes determining likeness in the framework of the classification of standards (OECD 1994):

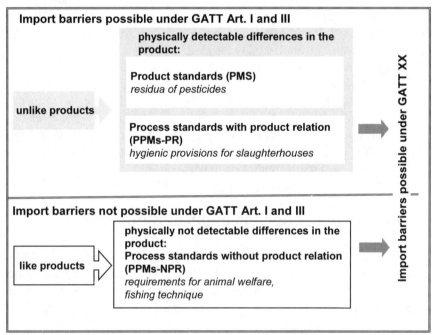

Source: OECD, 1994.

Fig. 1. Process and product standards

All standards with a physical and detectable impact on the final product (product measures (PMs), and process measures that are product-related (PPMs-PR)) may differentiate products into "unlike products" and thereby barriers can be compatible with GATT Articles I and III. Additional criteria ensure that such NTBs are not implemented arbitrarily and that least-trade distorting instruments are chosen. On the contrary, standards without any physical and observable impact (process measures that are not product-related (PPMs-NPR)) belong to the second category, namely leading to "like products". Therefore, these standards are not allowed to be applied to imports under Article I and III. As Article XX on general exemptions does not differentiate between like or unlike products for both types of standards, barriers could be justifiable in order to protect inter alia human, animal and plant health and life.[1]

A famous case referring to Article I and III are the two parts of the tuna-dolphin case of 1991 and 1994 on the ban of the United States (US) on Mexican tuna and on tuna originating from intermediary trading partners. The import ban was based on the requirement of the US to use a certain domestic fishing technique leading to less harm for the dolphin population. In both cases the panel interpreted the fishing technique as PPM-NPR and consequently, the US and Mexican tuna as "like" product. The US import ban therefore had to be abolished (case DS21/R-359).

An important exception is the product attribute "origin" that belongs clearly into the category of having no physical product impact. Nevertheless, rules of origin have been traditionally addressed by the WTO in the "Agreement on rules of origin" and the "Agreement on trade-related aspects of property rights" (TRIPS). According to these Agreements the differenttiation of products due to their origin is possible and numerous rules to enforce such a differentiation exist.

3   Time after the Adoption of the SPS-Agreement

The SPS-Agreement was adopted in 1994 and is comparable to the TBT-Agreement but consists of rules specifically for food matters and has some stricter provisions. It defines its scope of coverage by defining as SPS measures (Annex 1)

".. all laws, decrees, regulations, requirements and procedures, including, inter alia, end product criteria; processes and production methods; testing; inspection, certification and approval procedures, quarantine treatments including relevant requirements associated with the transport of animals or plants, or with the

---

[1] This argument was used by the US in the Turtle-Shrimps Case to justify the US import ban on shrimps of some Caribbean and other countries like Thailand and Malaysia as their fishing technique was deemed to be dangerous to sea turtles. Whereas the panel report had rejected the extraterritorial use of Article XX the Appellate Body in the contrary stressed that for moving species an extraterritorial application is not only allowed but even necessary. However, the ban was condemned due to discriminating effects of this specific measure. See DS58/RW and DS58/AB/R.

materials necessary for their survival during transport; provisions on relevant statistical methods, sampling procedures and methods of risk assessment; and packaging and labelling requirements directly related to food safety."

Hereby, product and process standards are mentioned but it is not explicitly stated whether these refer as well to PPMs without product relation. This is usually denied in several studies (James 2000) and is be empirically shown by the previously described outcome of respective cases (see Table 1).

3.1 The Provisions for National Flexibility

As key areas of the SPS-Agreement, the following issues will be discussed: (1) the accepted level of safety to be applied on imports by NTBs, and (2) the specific NTB to be implemented. For both areas the existing provisions and the given scope for national flexibility are described.

(1) Regarding the accepted safety level, the SPS-Agreement grants the general right to each member to implement such safety measures that are appropriate for achieving a chosen safety level in its territory (Article 2).

- To avoid trade distortion, harmonization is targeted as a key objective (Article 3). It is recommended to base national measures on international standards, guidelines and recommendations as far as they exist (Article 3.1). Such international standards are deemed to be necessary to protect human, animal or plant health (Article 3.2) repeating the general objectives of GATT Article XX. The resulting safety level can be interpreted as accepted by the WTO and therefore is not challengeable. The concrete *international standards* and guidelines are determined by a given catalogue of relevant institutions that are developing standards (Annex A 2-3): for food safety the Codex Alimentarius is the responsible institution jointly founded by FAO and WHO in 1964.[2] Codex standards cover for example maximum residua levels for antibiotics in pork or hormones in beef. This list of standards makes the SPS-Agreement different from the TBT-Agreement, where only the criteria for accepted standard-setting organisations are defined but no explicit list of organizations is given. Therefore, the SPS-Agreement can be interpreted as being stricter and in some disputes the defendant tries

---

[2] For standards related to other issues other Organisations are defined as responsible: for animal health the International Office of Epizootics for plant health the Secretariat of the International Plant Protection Convention (SPS-Agreement Annex A 2-3).

to base the respective barrier on the TBT-Agreement whereas the complaining party is using the SPS-Agreement.

Potential for national sovereign policy interms of deviating from these standards are related to the submission of a risk assessment to justify standards that are stricter than the Codex standards (Art. 3.3). The provisional implementation of stricter standards is possible even if scientific evidence to justify them is insufficient (Art. 5.7); however a risk assessment must be submitted at a later stage. This option is discussed intensively in the context of the precautionary principle. Most often Article 5.7 is not characterised as precautionary principle due to its terminal limitation and the need for scientific risk assessment at a later date (Gutpa 2000; Scott and Vos 2001). The specific requirements on risk assessments cover the criteria to be considered for a correct assessment, such as taking into account all relevant sampling methods (Art. 5). Only traditional risk dimensions like probability and damage amount are accepted as arguments, either quantitatively by figures or qualitatively by description (Annex A 4). The evaluation whether the submitted risk assessment is sufficient is the dominant argument in SPS disputes. The time period for filing the assessment subsequently is defined as "reasonable" and open to negotiations in the dispute procedure.

(2) Related to the choice of a specific NTB, a core rule of the WTO is to consider a minimal trade effect:

- Least-trade distortion is expressed in the SPS-Agreement as requirement to minimize trade effects (Articles 5.4, 5.6). As no measures are predetermined as being least-trade distorting, some general GATT principles have to be consulted to obtain information on what degree of trade restriction could be accepted. According to GATT Article XI no quantitative import restrictions are allowed and thus, import bans can be seen as the most problematic NTBs. An instrument often recommended as being very market-oriented, not trade distorting and an effective way to differentiate between product qualities is a label. Only few explicit provisions on labels can be found in the agreements. The TBT-Agreement is covering general packaging and label requirements, and the SPS-Agreement is addressing such issues when related to food (Annex A 1). In principle the same limitations as for standards are valid for labels. Therefore, no mandatory label for process standards without physical product impact is accepted as NTB whereas voluntary label are WTO conform (Josling et al. 2003). Related to accepted standards having a physical effect even mandatory labelling would be WTO compatible. For such label, harmonization is targeted. International standards for label that have been developed by Codex Alimentarius are recommended to aim at harmonization (e.g. STAN Serial of Codex, see Codex 2003). Hereby free trade is ensured and additionally the risk of abuse and the information overload for the consumer resulting from an intransparent variability of labels are reduced.[3]

- National scope to choose instruments is covered by the criterion of feasibility of NTBs and the principle of equivalence. The strict rule of using always the least trade-distorting measures is supplemented by additional criteria: the evaluation of implemented NTBs considers the technical and economic feasibility compared to alternative NTBs (Art. 5.6). The principle of equivalence can be understood as an alternative to the detailed harmonization of national food safety approaches. Equivalence means the acceptance of different instruments that achieve identical safety levels. This principle is recommended by allocating the burden of proof to the exporting country (Article 4) which has to convince its trading partner that the own safety instrument ensure the safety level of the importing country. The concrete implementation is realised by conformity assessments, i.e. the technical procedure to declare equivalence. Such procedures cover means to verify and document conformity, e.g. the intensity of inspections or the definition of critical levels of contamination (Josling et al. 2003). This granted possibility to maintain the national instrument is factually very rarely implemented. One reason is that the importing partner has to accept the equivalent performance. Very few bilateral agreements exist which are defining either minimum food standards and thereby are comparable to WTO rules or have to negotiate laboriously technical details (Rudloff and Simons 2004).[4] Finally, labelling offers some flexibility: a way out of harmonizing product labels can be the use of voluntary or private labels. These are not restricted or even not addressed by WTO. Therefore, private labels could be supported by accompanied public control procedures to increase effectiveness. There should be no public subsidies paid (e.g. for certification) because that could make private and voluntary

---

[3] A precedence became the „Sardine Case" between Peru and the EU. The EU restrictted the marketing to just one certain sardine specie under the term "sardines". Thereby sardines from Peru were excluded from market access. As the existing marketing standard of the Codex Committee referred to is applicable to a set of different species (Stan serie on packing and marketing requirements: Stan 94-1981, rev. 1-1995 and Stan 1-1985, rev. 3-1999) the EU's prohibition was condemned.

[4] An extraordinary example for a comprehensive Equivalent Agreement is Annex IV of the "EU-Chile Association Agreement" where detailed procedural elements such as inspection methods are ruled (EU-CHILE ASSOCIATION AGREEMENT 2002).

label challengeable either under the Agreement on Subsidies and Countervailing Measures or the Agreement on Agriculture (Rudloff 2003). For quality aspects instead of safety issues, such as cholesterol in food, more flexibility exists. Some general guidelines of the Codex Alimentarius exist without having the binding character of standards for the labelling of safety aspects. As no reference is made under the SPS-Agreement for food quality, harmonisation is not commanded for respective labels. This leads to flexibility on the one hand but to huge intransparency for the consumer on the other hand (Caswell 1997).

3.2 Survey on Disputes

General Overview on Food-related Cases

After the formal foundation of the WTO in 1994 and the adopted reform of the dispute procedure, 328 cases were opened formally by requesting consultations. According to the reform 1995 the procedures have been strengthened leading to more actually concluded disputes. The following Table 2 indicates the factual relevance of conflicts between WTO members due to SPS issues compared to other conflict areas. The Table covers opened cases referring to different WTO-Agreements since 1995. Opened disputes are covering all formally announced disputes starting with the status of request on consultation:

Table 2. Empirical relevance of WTO disputes on NTBs (January 1995 – March 2005)

All cases	328
Reference to Agreement on Agriculture	55
Reference to Sanitary and Phytosanitary Agreement	30
Reference to Agreement on Technical Barriers to Trade	32

Source: Own calculation on basis of the published cases at http://www.wto.org/ english /tratop_e/dispu_e/dispu_subjects_index_e.htm (March 2005).

Out of the 328 cases 55 are referring to the Agreement on Agriculture and altogether 62 are referring to NTBs either under the SPS or the TBT-Agreement showing the increasing relevance of conflicts on NTBs.

Since the adoption of the SPS-Agreement in 1994, thirty formal cases on food safety have been opened till today (Table 3):

Table 3. Overview on SPS disputes (January 1995 – March 2005)

Basis for cases opened since 1995	Numbers
All SPS cases	30
Still active panels	5
Pending consultations	13
Mutually agreed solutions	5
Decided cases	5 (+ Asbestos)[5]
Cases with adopted reports (panel and appellate body)	5 (+ Asbestos)
Implementation of findings	3 (+ Asbestos)
Sanctions	2

Source: Own calculation on basis of the published cases at http://www.wto.org/english/tratop_e/dispu_e/dispu_subjects_index_e.htm (October 2004).

Nearly half of the decided cases were solved before they entered into all dispute stages. Therefore, a majority of the cases is not ending up in a judgement of the responsible WTO bodies. Formally announced bilateral compromises are mutually agreed solutions which account just for five cases. Additionally, other cases have been suspended without any formal final decision that may be caused by an informal consensus between the parties. This relevance of bilateral solutions demonstrates the self-enforcing power of the dispute settlement procedure to motivate solutions without awaiting formal findings.

Involvement of Developing Countries in Disputes

The provisions of special and differentiated treatment is an overall rule for all WTO agreements aiming at considering the specific situation of developing countries as integrated part of all WTO rules. Regarding the SPS-Agreement, this principle grants longer phasing-in periods for

---

[5] The Asbestos Case is only formally referring to the SPS-Agreement but not addressing any food-related matter. Therefore this case will not be covered by the following analysis.

implementing new standards, the possibility for overall exceptions from duties and recommends assistance to join relevant organisations such as the Codex Alimentarius Commission (Art. 10). As developing countries often are underrepresented at such meetings due to lack of financial and human resources the Trust Fund offers support to visit the regular Codex meetings. Hereby, representatives of developing countries may actually contribute to the definition of standards that afterwards will become the harmonized ones under SPS.

Table 4. Involvement of developing countries in food disputes (March 2005) [1]

Involvement of developing countries in SPS-disputes	Least developed countries [2]	Low income countries [3]	
		Low income	Lower Middle income
... as defendant	0	India India	Egypt Turkey Turkey
... as complainant	0	India Nicaragua	Philippines Philippines Thailand Ecuador

Source: Own calculation on basis of the published cases at http://www.wto.org /english/tratop_e/dispu_e/dispu_subjects_index_e.htm (March 2005).
Notes:
1) At WTO the affiliation to developing countries is based on self-declaration and has not been considered in the Table.
2) Least developed countries are classified according to the UN's Index 2003. According to this classification no least developed country was involved in disputes.
3) The income classification is based on the World Bank's Atlas approach (for 2004: low income = $765, lower middle income = $766 - $3,035).

More and more developing countries are involved in food disputes as both defending and complaining party what is indicated Table 4. In all opened thirty SPS cases low income and lower middle income countries account for eleven. So far no least developed country (LDC) has been involved in any food disputes but one case of non-food disputes exists. Half of the disputes take place between developing countries.

As all these cases are still at the very beginning of the overall dispute procedure, no formal reports have been published clarifying the underlying details. Thereby, the following analysis will focus only on cases among developed countries as these are the only concluded cases.

Findings of Disputes Related to National Flexibility

The following five cases were closed and serve as basis for the analysis on granted scope for national sovereignty. Only the two Hormone Cases are directly linked to food safety aiming at human health. The others are targeting at plant health (Fruit Case and Apple Case) or animal health (Salmon Case) and therefore harmonization is based on other international standards than those of Codex Alimentarius.

(1) The *Salmon Case*: Canada accused Australia for having implemented an import ban on salmon that is not fulfilling Australian heating treatment requirements (WT/DS18).

(2–3) The two *Hormone Cases* in which both the US and Canada complained about the European import ban on meat produced with growth hormones (WT/DS26 and WT/DS48).

(4) The *Fruit Case*: the US complained against Japan applying domestic quarantine requirements on imports of certain fruit products and nuts in order to avoid the spread of codling moths (WT/DS76).[6]

(5) The *Apple Case* in which the US complained about the Japanese application of certain quarantine requirements on imports to avoid the spread of fire blight (WT/DS245).[7]

For the majority of these cases, the findings were made in favour of the complainant what is a general trend for all WTO disputes. Just two cases ended with the final institutional stage, i.e. retaliation: because the loosing parties failed in the implementation of the findings in terms of abolishing the measure at stake, penalty tariffs were applied by the complaining party.[8] This situation has appeared in both *Hormone Cases* that have not

---

[6] This insect is not dangerous for human health but destroys the harvest. The infection is depending on climate conditions leading to a differentiation of import requirements depending on the season.

[7] Fire blight is a plant disease not harmful for human health but hindering the mildewed products from being marketed.

[8] Using this option of penalties can be found very rarely when looking at all disputes. This can be explained by the reputation effect, i.e. losing international reputation due to not following the rules. Another reason is the fact that such trade reducing penalty tariffs are of disadvantage to both parties because even for the winning party welfare losses appear due to reduced imports (Hudec

been solved in terms of abolishing the condemned import ban and have remained in the status of keeping the ban while imposing penalty tariffs till today. Another case that led to the request of penalty tariffs is the Apple Case. A decision on granting penalties has been suspended until further notice. As the *Hormone Cases* were the first ones closed they are often used as precedence and are referred to in the other cases.

The following Figure 2 summarizes the relevant issues for national scope and indicates the findings of existing cases made in favour of national sovereignty.

(1) Related to the safety level, the core argument in all cases was the scientific justification for the chosen safety level (Article 3 and 5). In most cases the scientific assessment was rejected as inappropriate. Just in the two *Hormone Cases* the insufficient scientific evidence was accepted to justify the provisional establishment of the import ban according to Article 5.7. The granted period was 15 months. On the contrary, such option was rejected in the *Apple Case* as the scientific evidence was evaluated as being sufficient. In the *Fruit Case* the second condition for implementing Article 5.7. was evaluated as insufficient, namely that Japan failed in searching for all information available.

(2) Regarding the implemented instrument, in half of all cases the NTB at stake was accepted as the only feasible one compared to alternatives. Even the most trade distorting import ban in the two *Hormone Cases* and in the *Salmon Case* was accepted as the only technically feasible one compared to alternative measures such as process controls. Nevertheless, the ban was finally condemned in all cases but due to the missing risk assessment and not because of the trade distorting effect as such. In the *Fruit Case,* the panel accepted the testing methods required for imports as being the only feasible measures. But the subsequent appellate body rejected the argument as being formally not relevant for the findings. The principle of equivalence was not addressed in any of the cases.

1996). In the Hormone Cases both the US and Canada as complainants against the EU had difficulties in choosing the products on which they wanted to impose those penalty tariffs (Rudloff 2003).

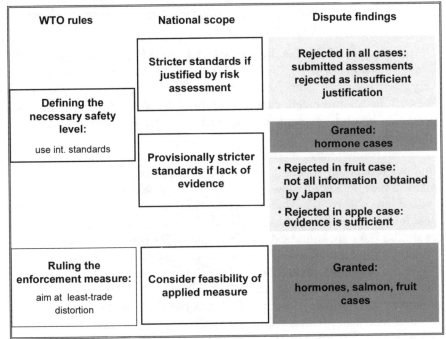

Source: Own composition.

Fig.2. Granted scope for a national food policy design

4 Conclusions

The analysis of existing WTO provisions has identified limited scope for national sovereignty. The existing scope can be different related to single aspects:

1. For the enforcement of domestic safety levels at the border, little scope exists. If international standards have been developed the only way out of harmonisation is the submission of a risk assessment, which is the most often used argument in disputes to reject a NTB at stake. The most important flexibility for the safety level is offered by allowing provisional measures if scientific evidence is insufficient. This flexibility is timely restricted as the risk assessment has to be submitted at a later date.

2. The largest scope for national action exists in the area of choosing a specific NTB. Feasibility can justify instruments that are not accepted as least trade distorting. In half of all cases the measure at stake was accepted due to this reason. Equivalence is suggested as facilitating

instruments but hardly used by countries due to necessary enormous bilateral bargaining efforts.

The analysis of the concluded cases has shown that even the existing windows for national flexibility is limited by strict criteria leading to its very rare use. The dominance of scientifically based food policies is stressed in all presented cases. Thereby the WTO dispute bodies are becoming involved in evaluating scientific soundness instead of pure trade impacts.[9]

These results must be relativised in several ways: first, the WTO findings reflect only conflicts on internationally existing standards. But for many issues so far no standards have been developed and the standard setting process of the Codex Commission is lengthy. For conflicting positions on such issues the existing findings may only be relevant as far as similar risks are addressed for which analogies could be drawn. Like for Melengestrolacetat as one of the six hormones at stake in the Hormone Cases for which the Codex Commission had not developed a standard but the dispute bodies derived some conclusions (Rudloff 2003). Second, voluntary and private standards that are not covered by SPS rules are gaining increasing relevance, which is also true in the case of labels. For these standards flexible bilateral solutions are negotiable and not addressed by the WTO. And finally, all described findings are only related to the question of implementing stricter standards than existing international ones. Thereby the dispute results imply that existing international standards function as maximum standards as stricter standards never were accepted. Deviating from this requirement can only be followed by the very final mean to accept sanctions. This is in fact an institutionalised option in the WTO framework but the most rigid one. The *Hormone Cases* are the only ones where the status of the remaining import ban and the reacting sanction tariffs have been held up now for six years.[10] Hereby, the *Hormone Cases* symbolise the principal restriction of the global ruling frame when large differences on national policy objectives exist. But deviating from existing standards in the other direction, i.e. establishing weaker standards than international ones, has never been part of any dispute so far. Whether there is actual scope for undermining these standards is a question to be covered by empirical analysis of bilateral arrangements.

---

[9] The Appellate Body in the *"Hormone Case"* emphasised that the evaluation of the scientific quality could not be WTO's tasks (WT/DS/48, par.187).

[10] The sum of about 120 million $ is imposed as penalty tariffs on European products imported to US and Canada per year (Rudloff 2003).

A final issue relativising the relevance of the described findings is the fact that only the minority of conflicts actually is reaching the stage of a formal dispute. Therefore, another empirical question would be the analysis of arrangements taking place prior ever starting a dispute (see Henson in this proceeding).

References

Bagwell K, Staiger R (2002) The economics of the world trading system. MIT Press, Massachusetts

Caswell J (1997) Uses of food labeling regulations In: OECD working papers, vol.v (1997), no.100, Paris

Codex Alimentarius (2003) Procedural Manual 13th edn Rome, available at ftp://ftpfaoorg/codex/PM/Manual13epdf (Dec 2004)

EU-Chile Association (2002) Agreement on sanitary and phytosanitary as Annex IV, measures applicable to trade in animals and animal products, plants, plant products and other goods and animal welfare Available at http://europa euint/comm/trade/issues/bilateral/countries/chile/euchlagr_enhtm (Dec 2005)

Gandolfo G (1998) International trade theory and policy. 2nd edn Springer Verlag, Heidelberg

Gupta A (2000) Governing trade in genetically modified organisms: the Cartagena protocol on biosafety. In: Environment 42:23-33

Hudec R (1996) The GATT/WTO dispute settlement process: Can it reconcile trade rules and environmental needs? In: Wolfrum R (ed) Enforcing environmental standards: economic mechanisms as viable means? Beiträge zum ausländischen öffentlichen Recht und Völkerrecht, Bd 125, Duncker & Humblot, Berlin

James S (2000) An economic analysis of food safety issues following the SPS agreement: lessons from the hormone dispute. Centre for International Economic Studies. Policy discussion paper 0005, Adelaide

Josling T, Roberts D, Orden D (2003) Food regulation and trade, toward a safe and open global system. Institute for International Economics, Washington DC

OECD (1994) Trade and environment. Process and production methods. Organization for Economic Co-operation and Development, Paris

Rudloff B (2003) National concerns about food safety and international trade - An economic assessment of potential conflicts about food safety matters with an application to the "Hormone - Beef Case" (In German) PhD Thesis, University of Bonn In: Europäische Hochschulschriften, Peter Lang Verlag, Reihe V, Vol 2942, Frankfurt, 309 Seiten, ISBN 3-631-39777-1

Rudloff B, Simons J (2004) Comparing EU Free trade agreements - Sanitary and phytosanitary measures Prepared for ECDPM (InBrief), available at http://www.ecdpm.org /Web_ECDPM/Web/Content/Navigationnsf/index.htm

Scott J, Vos E (2001) The juridification of uncertainty: Observations on the ambivalence of the precautionary principle within the EU and the WTO. In: Dehousse R, Joerges Ch (eds) Good governance in an integrated market. Oxford University Press, Oxford

Unnevehr L, Roberts D (2003) food safety and quality: regulations, trade, and the WTO. Invited paper presented at the International Conference "Agricultural policy reform and the WTO: where are we heading?" Capri (Italy), June 23-26 2003

World Bank (2005) Food safety and agricultutural health standards: challenges and opportunities for developing country exports. Report No 31207: Poverty Reduction and Economic Management Trade Unit and Agriculture and Rural Development Department. World Bank, Washington DC

Cases

GATT: "Feed protein Case" Support domestic feed proteins – USA against EC, Case BISD 25S/49, March 1978

GATT: "Wood Case" Tariffs on certain woods – Canada against Japan, Case BISD 36S/167, July 1989

GATT: "Tuna Case 1" Restrictions on tuna import - Mexico against USA, Case DS21/R-359, 1991

WTO (1995) Agreement on agriculture. Available at http://www.wto.org/english/docs_e/legal_e/14-ag.pdf (December 2005)

WTO (1995) Agreement on rules of origin. Available at http://www.wto.org/english/docs_e/legal_e/22-roo.pdf (December 2005)

WTO (1995) Agreement on technical barriers to trade. Available at http://www.wto.org /english/docs_e/legal_e/17-tbt.pdf (December 2005)

WTO (1995) Agreement on the application of sanitary and phytosanitary measures Available at http://www.wto.org/english/docs_e/legal_e/15-sps.pdf (December 2005)

WTO (1995) Agreement on the understanding on rules and procedures governing the settlement of Disputes. Available at http://www.wto.org/english/docs_e/legal_e/28-dsu.pdf (December 2005)

WTO: "Salmon Case" Measures affecting the importation of Salmon-Canada against Australia, Case WT/DS18, June 1998

WTO: "Hormone Case 1" Measures concerning meat and meat products – USA against EC, Case WT/DS26, August 1997

WTO: "Hormone Case 2" Measures concerning meat and meat products – Canada against EC, Case WT/DS48, August 1997

WTO: "Sea turtle case" Import prohibition on certain shrimps and shrimp products – United States against Carribean States, Case WT/DS58, May 1998

WTO: "Fruit and Vegetable Case" Measures affecting agricultural products– USA against Japan, Case WT/DS76, October 1998

WTO: "Asbestos Case" Measures affecting asbestos and products containing asbestos – Canada against EC, Case WT/DS135, September 2000

WTO: "Sardine Case" Trade description of sardines - Peru against EC, Case WT/DS231, May 2002

WTO: "Apple Case" Measures affecting the importation of apples – USA against Japan, Case WT/DS245, July 2003

WTO: "Batteries Case" Request for consultations - Bangladesh against India, Case WT/DS306, January 2004

Contributors

Arnab K. Basu is Associate Professor of Economics and Public Policy at the College of William and Mary, Williamsburg, and a Senior Fellow at the Center for Development Research (ZEF) at the University of Bonn. His fields of interest include development and international economics.

Maria Cristina DM Carambas is an Economist with the Ontario Ministry of Natural Resources in Canada. She previously worked at the Center for Development Research of the University of Bonn as a Senior Researcher after obtaining her doctoral degree from the University of Giessen in Germany.

Sayan Chakrabarty is a Postdoctoral Fellow at the Center for Development Research of the University of Bonn, Germany. He obtained his doctoral degree in Economics from the University of Giessen, Germany. His research interest is in the area of empirical microeconomics and development economics.

Nancy H. Chau is Associate Professor at the Department of Applied Economics and Management at Cornell University, Ithaca, and a Senior Fellow at the Center for Development Research (ZEF) at the University of Bonn. Her fields of interest include regional, international and development economics.

Ahmed Farouk Ghoneim is Associate Professor, Faculty of Economic and Political Science, Cairo University, and a Research Associate at the Economic Research Forum for Arab Countries, Iran and Turkey (ERF). He holds a Ph.D. in Economics and his research interest is in trade policy.

Ulrike Grote is Senior Research Fellow at the Center for Development Research (ZEF), University of Bonn. Before that, she worked at the OECD and the Asian Development Bank. Her research focuses especially on international trade, environmental economics and social standards.

Spencer Henson is Professor in the Department of Food, Agricultural and Resource Economics at the University of Guelph, Canada. His research predominantly focuses on the impact of food safety and quality standards on developing countries and the implications for small-scale producers.

Robert L. Hicks is Associate Professor of Economics at the College of William and Mary, Williamsburg. His research includes econometric approaches for measuring consumer preferences for environmental goods, environmental valuation, and the optimal management of natural resources.

Stéphan Marette is a Research Fellow at the Department of Economics at INRA, France's National Institute for Agricultural Research. His research interests include product quality, labeling, liability and food safety.

Caroline L. Noblet is Research Associate / Lecturer for the Department of Resource Economics and Policy, and the Margaret Chase Smith Policy Center at The University of Maine. Her research interests include environmental education, environmental psychology / marketing and community development.

David Orden is Professor of Agricultural and Applied economics, Virginia Polytechnic Institute and State University, Blacksburg, Virginia, USA and Senior Research Fellow at the International Food Policy Research Institute, Washington D.C., USA. He is engaged in active research and public policy education programs on the economics and political economy of domestic support policies, international trade negotiations, and technical barriers to trade.

Everett Peterson is Associate Professor of Agricultural and Applied Economics, Virginia Polytechnic Institute and State University, Blacksburg, Virginia, USA and a Research Fellow of the Global Trade Analysis Project, Purdue University, West Lafayette, Indiana. His interests are in economic modeling with focus on firm behavior, marketing margins and other transaction costs in general equilibrium, as well as trade policy and technical barriers to trade.

Jonathan Rubin is Associate Professor in the Department of Resource Economics and Policy, and the Interim Director of the Margaret Chase Smith Policy Center at The University of Maine. He specializes in environmental economics and transportation energy policy.

Bettina Rudloff is Assistant Professor at the Institute for Food and Resource Economics of the University of Bonn, Germany. She is also a visiting lecturer at the European Institute for Public Administration (EIPA) in Maastricht, The Netherlands. Her fields of research are food safety and consumer protection, and governance issues.

Mario F. Teisl is Associate Professor in the Department of Resource Economics and Policy at The University of Maine. He specializes in environmental and health economics.